COASTAL GEOMORPHOLOGY: AN INTRODUCTION

Coastal Geomorphology: An Introduction

Editor:

Navodita Bhatnagar

Delve Publishing

www.delvepublishing.com

Coastal Geomorphology: An Introduction

Editor: **Navodita Bhatnagar**

708 3rd Avenue, 6th Floor
New York, NY 10017
United States of America
www.delvepublishing.com

ISBN: 978-1-68095-823-2

Library of Congress Control Number: 2016951565

About the Editor

Navodita Bhatnagar

Navodita Bhatnagar finished her masters in technology and currently pursuing her PhD in environmental sciences from Institute of Technology, Carlow, Ireland. She's a keen enthusiast of environment protection and management. An alumunus of Indo-German centre for sustainibility which is essentially a collaborative interface to gather scientists in the field of environment sciences to come together and work for the common cause which is environment protection.

Contents

List of Contributors ... xi

Preface ... *xiii*

Chapter 1 Coastal Erosion Studies—A Review ... 1

Abstract ... 1

1. Introduction ... 1

Conclusions ... 3

References ... 4

Chapter 2 Effects of the Coastal Environment on Well-Being ... 5

1. Introduction ... 5
2. Methodology ... 6
3. Results ... 6
4. Discussion ... 8

Conclusion ... 11

References ... 13

Chapter 3 Geomorphological Coastal Classifications After Natural Processes and Human Disturbance ... 15

1. Classifications After Natural Processes ... 15
2. Coastal Classification After
3. Human-Disturbance ... 15

Conclusions ... 16

References ... 16

Chapter 4 Investigation of Integrated Coastal Management Planning Model: Problems and Solution Alternatives-Coasts of Sinop ... 19

1. Introduction ... 19
2. Materials and Method ... 20
3. Results and Discussion ... 21

References ... 29

Chapter 5 Studying Heavy Metals In Sediments Layers Along Selected Sites on the Lebanese Coast ... 31

Abstract ... 31

1. Introduction ... 31
2. Methodology ... 323.
3. Results ... 33

References ... 53

Chapter 6 Beach Erosion-Coastal Protection and Rehabilitation Related to Climate Change ... 55

1 .Introduction ... 55

References ... 56

Chapter 7 Yellowfin Seabream, a Target Species for Monitoring of Coastal Pollutants ... 57

1. Fish Species For Eco-Toxicological Studies ... 57
2. Yellowfin Seabream, Suitable Marine Fish Species ... 58

References ... 58

Chapter 8 **Macrobenthic Community Structure as a Bio-Indicator for the Assessment of Coastal Water Pollution In Greater Noakhali-Bangladesh** **59**

1. Introduction 59
2. Materials And Methods 59
3. Results And Discussion 60

Conclusion 66

References 66

Chapter 9 **A New Method for Rapid Treatment and Management of Coastal Oil Pollution** **69**

1. Introduction 69
2. Materials and Methods 71
3. Method 72
4. Discussion 72

Conclusion 73

References 73

Chapter 10 **Remote Sensing of the Coastal Ecosystems** **75**

1. Introduction 75
2. Perspectives 76

References 77

Chapter 11 **The Role of Remote Sensing and Gis for Locating Suitable Mangrove Plantation Sites Along the Southern Saudi Arabian Red Sea Coast** **79**

Abstract 79

1. Introduction 79
2. Geographical Location of the Study Area 80
3. Materials and Methods 80
4. Results and Discussion 82

Conclusions 83

Acknowledgements 83

References 84

Chapter 12 **Coastal Vulnerability and Shoreline Changes for Southern Tip of India-Remote Sensing and Gis Approach** **85**

Abstract 85

1. Introduction 85
2. Study Area and Geomorphology 86
3. Materials and Methods 88
4. Results 89
5. Discussion 92

Conclusion 94

Acknowledgement 95

References 95

Chapter 13 **The Importance of Environmental Safety Evaluation of Beaches** **97**

1. Introduction 97

Conclusion 98

References 98

Chapter 14 **Nutritional Security, Healthy Marine Ecosystems and Value Added Priorities for Developing Coastal Countries** **99**

References 100

Chapter 15 **The Scope of Marine Spatial Planning and Integrated Coastal Zone Management: New Challenges for the Future** **101**

1. Editorial 101

References 102

Chapter 16 **Effects of Land Use Changes on Ecosystem Services Value Provided by Coastal Wetlands: Recent and Future Landscape Scenarios** **103**

1. Introduction 103
2. Methods 104
3. Results 106
4. Discussion and Conclusions 108

Acknowledgements 109

References 109

Abstract 111

Chapter 17 **Coastal Flooding in the Solent: An Integrated Analysis of Defences and Inundation** **111**

1. Introduction 111
2. The Solent Case-Study Region And Initial Assessment Of Coastal Flood Risk 112
3. Methods 114
4. Results 120
5. Discussion 125

Conclusions 126

Acknowledgments 126

References 126

Index **131**

List of Contributors

Durusoju Hari Prasad
Department of Civil Engineering, JNTU, Hyderabad, India

Nandyala Darga Kumar
Department of Civil Engineering, JNTU, Hyderabad, India

Chenchen Peng
Graduate School of Maritime Sciences, Kobe University, Japan

Kazuo Yamashita
Graduate School of Maritime Sciences, Kobe University, Japan

Eiichi Kobayashi
Graduate School of Maritime Sciences, Kobe University, Japan

Alcantara-Carrio J
Oceanographic Institute, University of Sao Paulo. Pca Oceanographic, 191, Butanta, SP, 05508 900, Brazil
Institute of Environment Science & Marina, Catholic University of Valencia, 94 Guillem de Castro, 46001 Valencia, Spain
Area of Marine Sciences, University EAFIT, AA 3300, Medellín, Colombia

Fontán Bouzas A
CIMA/University of Algarve, Building 7, Campus Gambel Faro, Faro 8005-139, Portugal

Albarracin S
Institute of Environment Science & Marina, Catholic University of Valencia, 94 Guillem de Castro, 46001 Valencia, Spain

Correa ID
Area of Marine Sciences, University EAFIT, AA 3300, Medellín, Colombia

Montoya Montes I
Oceanographic Institute, University of Sao Paulo. Pca Oceanographic, 191, Butanta, SP, 05508 900, Brazil

Mahiques M
Oceanographic Institute, University of Sao Paulo. Pca Oceanographic, 191, Butanta, SP, 05508 900, Brazil

Oylum Gökkurt Baki
Environmental Engineering Department, Faculty of Engineering and Architecture, Sinop University, Turkey

Osman Nuri Ergun
Environmental Engineering Department, Faculty of Engineering, Ondokuz Mayis University, Samsun, Turkey

Nadine NASSIF

Department of Environment, Agricultural Engineer, Lebanese University, Dekwanek, Lebanon

Ziad SAADE

Department of Environment, Agricultural Engineer, Lebanese University, Dekwanek, Lebanon
Ramessur RT
Department of Chemistry, Faculty of Science, University of Mauritius, Reduit

Aliakbar Hedayati*
Faculty of Fisheries and Environment, Gorgan University of Agricultural Sciences and Natural Resources, Gorgan, Iran

Md. Jahangir Sarker
Department of Fisheries and Marine Science, Noakhali Science and Technology University, Noakhali-3814, Bangladesh

Mehedi Hasan Tanmay
Department of Fisheries and Marine Science, Noakhali Science and Technology University, Noakhali-3814, Bangladesh

Farhana Rahman
Department of Fisheries and Marine Science, Noakhali Science and Technology University, Noakhali-3814, Bangladesh

Md. Shamsul Alam Patwary
Department of Fisheries and Marine Science, Noakhali Science and Technology University, Noakhali-3814, Bangladesh

Nazmun Nahar Rima
Department of Fisheries and Marine Science, Noakhali Science and Technology University, Noakhali-3814, Bangladesh

Mu-Chien Lin
Department of Maritime Police, Central Police University, Taoyuan City, Taiwan

Victor Klemas
College of Earth Ocean and Environment University of Delaware Newark, DE 19716, U.S.A.

Ibrahim H. Elsebaie
Civil Engineering Department, College of Engineering, King Saud University, Riyadh, KSA

Ayman S. H. Aguib
Civil Engineering Department, College of Engineering, King Saud University, Riyadh, KSA

Dafer Al Garni
Civil Engineering Department, College of Engineering, King Saud University, Riyadh, KSA

V. Joevivek
Centre for GeoTechnology, Manonmaniam Sundaranar University, Tirunelveli 627102, India
Civil engineering department, Christ the king engineering college, Karamadai, Coimbatore 641104, India

S. Saravanan
Centre for GeoTechnology, Manonmaniam Sundaranar University, Tirunelveli 627102, India

N. Chandrasekar
Centre for GeoTechnology, Manonmaniam Sundaranar University, Tirunelveli 627102, India

Changchun Zhou
College of Geography and Tourism, Qufu Normal University, Rizhao 276826, Shandong, China

Pierre Failler
Senior Research Fellow Centre for the Economics and Management of Aquatic Resources (CEMARE) University of Portsmouth St. George's Building 141 High Street Portsmouth, PO1 2HY, United Kingdom

García-Sanabria Javier
Faculty of Marine and Environmental Sciences, University of Cádiz, Spain

Vera Camacho Valdez[1]
El Colegio de la Frontera Sur, Carretera Panamericana y Periférico Sur s/n, Barrio María Auxiliadora, San Cristóbal de las Casas, Chiapas, Mexico

Arturo Ruiz-Luna
Center for Food Research and Development, Sinaloa, Mexico

César Berlanga-Robles A
Center for Food Research and Development, Sinaloa, Mexico

Matthew P. Wadey
Faculty of Engineering and the Environment, University of Southampton, Southampton SO17 1BJ, UK

Robert J. Nicholls [1]
Faculty of Engineering and the Environment, University of Southampton, Southampton SO17 1BJ, UK

Craig Hutton
Geodata Institute, University of Southampton, Southampton SO17 1BJ, UK

Preface

Geomorphology, in simple words, is the study of landforms. These studies incorporate the in-depth understanding of various processes occurring on earth's surface in form of soil, water and air as the medium which directly or indirectly affects the landscapes. Present day landforms are a result of millions of years of geomorphological processes that continue to occur at their own pace now and in future. Few examples of natural geomorphological processes are soil erosion or deposition, melting of glaciers, volcanic eruptions, tsunamis, sea floods, landslides, and avalanche.

Coastal geomorphology focuses on history of formation of coasts, the features of coastal areas across the globe, eroding shorelines, processes involved in making the coasts as they are today like winds, ocean currents, and influence of human activities. A major proportion of world's population resides by the coasts and thus they are affected greatly by the anthropogenic activities. Shoreline recession has become an issue of concern in recent times which is occurring way faster than expected and almost over 70% of the beaches throughout the world are facing this problem. Chapters that deal with erosion of the coasts, coastal management, bioindicators for coastal pollutants and various techniques like remote sensing and GIS for the analysis of the coastal ecosystems have been included to provide a better understanding of present day situation. Studies pertaining to preservation of coastal integrity will be focused in this book.

Coasts are also an inevitable part of a country's economy in terms of tourism and water based trade. The economic and social aspects have also been looked into in the book. I hope the book provides the readers with a thorough understanding of the subject.

Editor
Navodita Bhatnagar

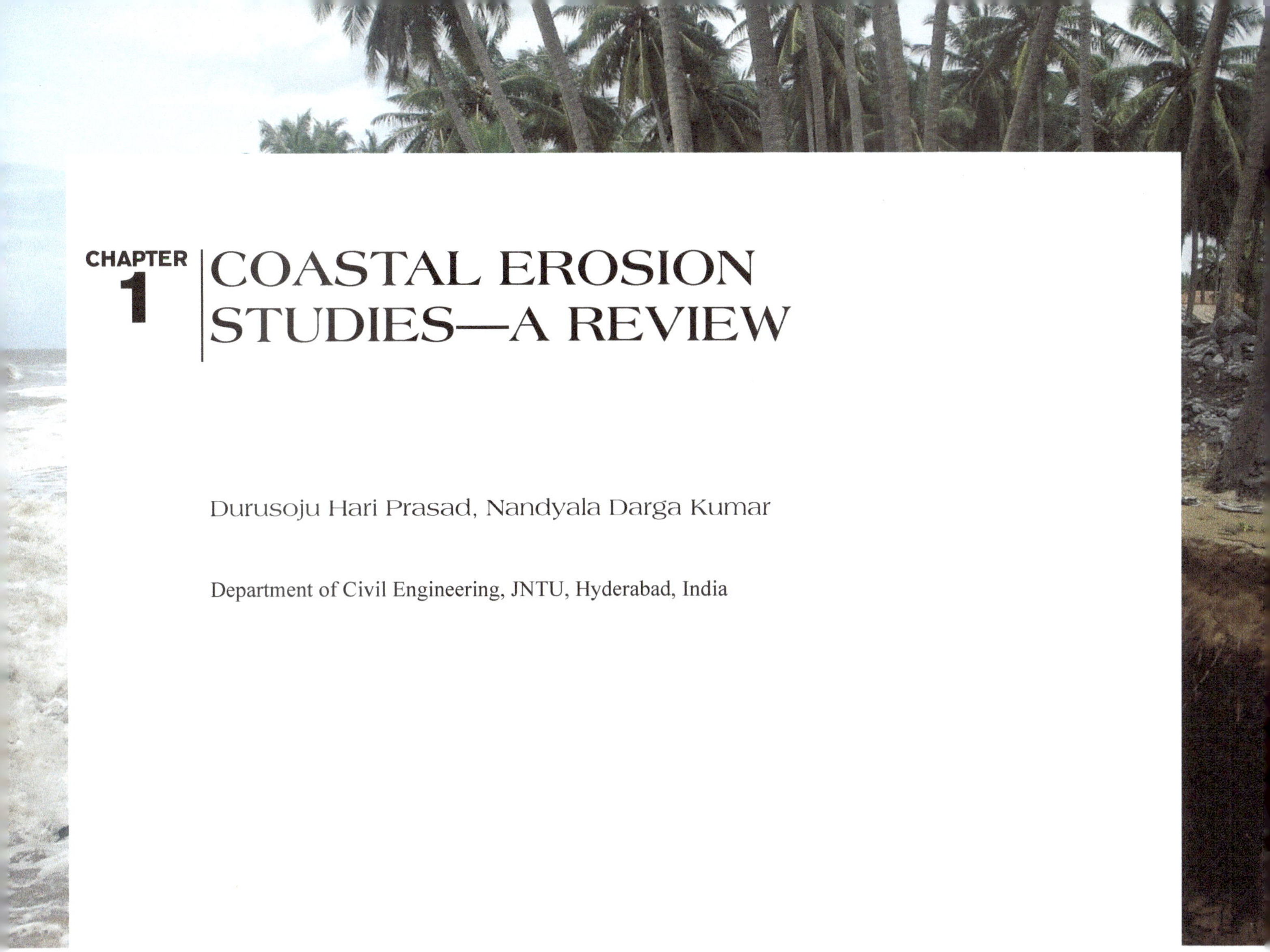

CHAPTER 1

COASTAL EROSION STUDIES—A REVIEW

Durusoju Hari Prasad, Nandyala Darga Kumar

Department of Civil Engineering, JNTU, Hyderabad, India

ABSTRACT

The land ward displacement of the shoreline caused by the forces of waves and currents is termed as Coastal Erosion. The Coastal areas have become more prone and vulnerable to natural and human made hazards which lead to Coastal Erosion. The Shoreline retreat is recognized as a burgeoning threat because of global climate change and other anthropogenic activities that alter the natural processes of sustaining beaches and coasts. Coastal Erosion mainly occurs when wind, waves and long shore currents move sand from shore and deposit it somewhere else. The sand can be moved to another beach, to the deeper ocean bottom, into an ocean trench or onto the landside of a dune. The removal of the sand from the sand sharing system results in permanent changes in beach shape and structure. The impact of the event is not seen immediately as in the case of Tsunami or Storm Surge but it is equally important when we consider loss of property. It generally takes months or years to note the impact of Erosion; therefore, this is generally classified as a "Long Term Coastal Hazard". The present paper attempts to describe a Review on Coastal Erosion Process, Parameters' affecting, and methodologies are adopted to identify the erosion and recommend a solution.

INTRODUCTION

Shoreline or coastline, the boundary between land and sea, keeps changing its shape and position continuously due to dynamic environmental conditions. Various developmental projects are made in shoreline areas, placing great pressure on it, leading to diverse coastal hazards like soil erosion, sea water intrusion, coral bleaching, shoreline change; etc. Considering Coastal erosion, it is a global problem affecting almost every country around the world having a coastline. It's a hazard effecting the shoreline or coastline pertaining to several changes in the climate, atmospheric disturbances and constant changes in the water bodies.

However, globally it is estimated that about 60% of the population is dwelling in the coastal environments. Although the coastal environment can retain some degree of natural character, increased human modification reduces the "naturalness" (Dahm, 2000). The accelerated release into the atmosphere of Carbon dioxide and other greenhouse gasses has resulted in a projected global warming of about 3°C by the year 2030 (Davis & Fitzgerald 2010) [1] .

This increase would be enough to raise the global sea level by as much as 5 m in a few centuries (Davis & Fitzgerald 2010), which is a short time in terms of human occupation of the coast. This phenomenon has the tendency to expose a significant proportion of vulnerable coastal areas to flooding

Citation: Prasad, D. and Kumar, N. (2014) Coastal Erosion Studies—A Review. International Journal of Geosciences, 5, 341-345. doi: 10.4236/ijg.2014.53033.

and destruction of habitats for migratory birds and other endangered species. Occurring in synchrony with these continual changes is the recession of shorelines, estimated to occur on coastal population and infrastructure of major economic and cultural importance in almost all coastal regions globally.

The location of the shoreline and changing position of this boundary through time are of elemental importance to coastal scientists, engineers and managers (Douglas & Crowell 2000). Relief and development agencies also depend on such information to facilitate development of effective measures to prevent, mitigate or manage disasters. Both sustainable coastal management and engineering design require information about where the shoreline is, where it has been in the past, and where it is predicted to be in the future. Such information is required in the design of coastal protection (Coastal Engineering Research Centre 1984), to calibrate and verify numerical models (Hanson et al. 1988), to assess sea-level rise (Leatherman 2001) and to develop hazard zones.

Remote Sensing helps to replace the conservative survey data by its rhythmic and less cost-effectiveness. Several studies using satellite data have proven its efficiency in understanding various coastal processes (Wagner et al. 1991; Ahmed and Neil 1994; Anbarasu et al., 1999; Makota et al., 2004; Mani Murali et al., 2009; Boutiba and Bouakline 2011) [2] . Space technologies have the capability to provide information over a large area on a repetitive basis, and therefore, it's very useful in identifying and monitoring various coastal features. Douglas, Crowell & Leatherman 1998, formulate policies to regulate coastal developments (National Research Council 1990), and to assist with legal property boundary definition (Morton & Speed 1998) as well as coastal research and monitoring (Smith & Jackson 1992).

Causes and Impacts of Coastal Erosion

A coastline is a complex series of interlinked physical systems in which both offshore and onshore processes are involved. Coastal Erosion is one of these physical processes, wearing away and redisturbing solid elements of the shoreline as well as sediment, normally by such natural forces as waves, tidal and littoral currents and deflation. The coastal sediments, together with those arising from inland erosion and transported seaward by rivers, are redistributed along the coast, providing material for dunes, beaches, marshes and reefs.

Coastal Erosion is usually the result of a combination of factors both natural and human induced operating on different scales. Erosion is defined as the encroachment of land by the sea after average over a period which is sufficiently long to eliminate the impacts of weather, storm events and local sediment dynamics (such as Sand Waves) [3] .

Coastal Erosion results in three different types of impact like: Loss of land with economical value (such as the beaches) or with ecological value; a specific mechanism is the collapse of properties located on the top of cliffs and dunes.

Models for Predicting Shoreline Change

A wide variety of numerical models and methods have been proposed to detect the shoreline change and also to model the change mathematically.

In many cases the parameters such as surf zone turbulence, effect of frequency, variations in directional spread and other wave transformations are also calculated and the solutions are formulated. Therefore some of the models are been briefly discussed in this paper.

Equilibrium Model

Shoreline Change is affected by a multitude of complex processes operating at a various time and length scales (Larson & Kraus, 1995). At shorter time scales on energetic coastlines wave dominated cross-shore transport, long shore transport gradients, wave set-up and storm surge are the dominant processes driving shoreline change. Coastal managers, scientists and engineers have long sought a robust and practical methodology for the prediction of shoreline change along sandy coastlines, over time-scales spanning several years to decades. The equilibrium model [4] , which at the present time come closest to satisfying theses requirements generally include a considerable level of empiricism and may be termed top-down or data-drive models. Probably the best known and most widely used example is the GENESIS model which is applicable to predicting generalized platform shoreline evolution for the special case where alongshore gradients in sediment transport dominate.

A variety of measures are used here to objectively assess model skill. The first is linear squared-correlation (r2) between the measured (x) and modeled (x_m) shoreline position. Whilst this method is useful for exploring correlations between the measurements and predictions it is possible that the series may be well correlated but have large residual. For this reason more rigorous comparative methods are also used, which compare the model residuals with a suitable baseline (x_b). The choice of baseline is somewhat arbitrary. Here, both a linear fit to the shoreline series and the prioir DLT10 model for shoreline position are used to assess improved model performance. The first comparative method, the Brier Skill Score (Sutherland and Soulsby, 2003), also has the advantage of considering measurement error Δx.

$$\text{BSS} = 1 - \sum\{|x - x_m| - \Delta x\}^2 / \sum(x - x_b)^2 \qquad (1)$$

AHP Model

The Analytical Hierarchy Process [5] (Saaty, 1980) is a multi-criteria decision method that solves decision making problems by ranking alternatives according to several criteria. The main steps in the method involve –

- Representing a decision making problem by organizing its criteria into a hierarchical structure.
- Evaluating the relative importance of the criteria (within a hierarchical tree) and then the alternatives with respect to each criteria on; this is done by constructing pair wise comparison.

To estimate coastal shoreline change, we generally apply simplified models and the associated scaling of uncertainties, taking into account the susceptibility of the coast to erode.

The simplified model of beach erosion is the formula suggested by the Eurosion (2004) project, which combines expected erosion due to sea level rise (Brunn, 1962) with shoreline change due to ongoing processes, as calculated from past shoreline observations

$$R_{future} = R_{historical} + S_{future} - S_{historical} / \tan(\alpha) \quad (2)$$

where tan(α) is the beach slope and S is the sea level rise. For cliff erosion we combine spatially continuous knowledge of past erosion, local knowledge of erosion rates calculated for a few representative sites and expert opinions. The use of these models is arguable, so we consider a scaling of uncertainties in potential future erosion instead of a single future shoreline position.

Model Using SPH Method

Smoothed Particle Hydrodynamics Method (SPH) [6] is a relatively new method for examining the propagation of highly non-linear breaking waves. The SPH method can be considered as computing the trajectories of particles of fluid which can interact according to the Navier-Stokes equations. An alternative view is that the fluid domain is represented by the nodal points that are scattered in space with no definable grid structure and move with the fluid. Each of these nodal points carries scalar information, density, pressure, velocity components, etc.

$$f(x) = \sum f_{jj} W(x - x_j) V_j \quad (3)$$

Taking the conservation of Mass and momentum written in particle form are taken into account (Monaghan, 1992), we finally derive an equation that relates the pressure in the fluid to the local density.

$$P = B\left[(p/p_o)^{\gamma} - 1\right] \quad (4)$$

The factor Gamma is taken as 7. This equation implies that the fluid is compressible.

In this way different equations are derived for other components such as velocity, viscosity, which also incorporates the numerical modeling of large and small turbulent eddies and the trajectories with which the wave strikes the shore. Finally, the derived equations are interpolated with the standard values and conclusions are drawn.

Model Formulating Sand Transport

Sand transport becomes crucial parameter while estimating the coastal erosion. Practical sand transport formulae for the coastal marine environment are generally semi-empirical formulae which can be classified as time-averaged, quasi-steady or semi steady. Based on approaches used for fluvial sediment transport, time averaged formulae predict sand transport at a timescale that is much longer than the wave period, using wave-averaged values of velocity and sand concentration. The bijker (1971) [7] formula is an example of a widely-used time-averaged transport formula where the total net transport is always in the direction of the mean current and the wave-related transport component is not taken into account. Thus, taking the above quantities into account the new formula is based on a modified version of the semi-unsteady "half cycle" concept originally proposed by Diajnia and watanabe (1992). In this concept, the wave-averaged total net sand transport rate as taking place in the oscillatory boundary layer is essentially described as the difference between the two gross amounts of sand transported during the positive "crest" half-cycle and during the negative "trough" half-cycle. Unsteady phase lag effects are taken into account. In addition to this the formula uses 1) bed shear stress rather than nearbed velocity as the main forcing parameter; 2) the phase lag effects are considered; 3) the effects of acceleration skewness are incorporated 4) it also covers graded sands and 5) the formula distinguishes between oscillatory flows and progressive surface waves.

Applications using GIS & Remote Sensing

The shoreline is one of the rapidly changing coastal landforms. Shorelines are the key element in coastal GIS and provide the most information on coastal landform dynamics. Therefore, accurate detection and frequent monitoring of shorelines are very essential to understand the coastal processes and dynamics of various coastal features. Nevertheless, the researchers and engineers have been studying the several levels of the increasing complexities of wave information from offshore to near shore, the role of different technology strategies have always been helpful in drawing important conclusions and also to make their job easier. The usage of Remote sensing devices, satellite data, topographical maps and electronic devices operating in the microwave frequency bands such as RADARS [8] are extensively used nowadays. Precisely, the Synthetic Aperture Radar (SAR) images are being used more often than ever before for geosciences applications in the moist tropics.

The airborne GEMS 1000 X-HH radar images acquired in 1972 during the RADAM Project was also used for evaluating coastal changes occurring over last three decades. In application perspective, orbital and airborne SAR data proved to be fundamental source of information for both geomorphological mapping monitoring coastal changes in moist tropical environments. At present remote sensing spatial resolution has achieved even 1 cm or better in the coastal zone. The current in-orbit satellites of oceans include NOAA series satellites, Landsatseries satellites, MOS-1, JERS-1, China meteorological satellites FY-N, SPOT, ERS, MODIS, Sea WiFS, [9] China resource satellite ZY-1, etc. By using large-scale and precision positioning macro image information provided by varies satellites, coastal scientist can monitor various coastal geomorphology types as well as their spatial distribution information about estuarine delta, tidal flat, coastal dunes, shell ridges, barrier lagoon, mangrove and coast coral reefs, etc. Image data can also be acquired from multi-spectral scanners or radar sensors aboard aircraft.

CONCLUSIONS

The review paper presents an overview of the significant work done in the geosciences field, wherein a new efficient three-dimensional model is introduced to calculate the water wave intermediate velocities for the advection and diffusion by imposing the divergence-free velocity field condition. The 3D model is shown to well resolve vertical velocity profiles of nonlinear wave deformation over a submerged bar.

The SPH technique, with its Lagrangian formulation, provides a methodology for the detailed examination of water vorticity. It is particularly suited to those cases where there is splash or flow separation.

Moreover, determination of extreme beach erosion involves integrating a five to six parameter probability density function of wave parameters (height, period, direction and storm duration), surge and timing. The wave details are often recorded well offshore or taken from global wave models well offshore where the hindcast wave parameters are considered free from near shore bathymetry that is coarsely represented or assumed deep water in these models due to their scale.

In addition to this, RADARSAT's capability to provide unique geomorphological and land cover mapping of the Coastal Plain and the integration of information provided by the orbital C-HH imagery with the airborne XHHSAR [10] has allowed detecting and mapping coastal changes related to the shoreline retreat and accretion, during the 26-year period of both SAR acquisitions. Sedimentary dynamic, tidal current, wave action and estuarine and tidal channel displacements have played an important role in controlling these coastal changes. Based on results obtained from SAR data analysis, we can conclude that SAR imageries have provided valuable, rapid and accurate information on coastal features recognition, coastal land-use assessment, monitoring of shoreline changes and base maps' updating, which are basic components for an integrated management program in this coastal zone. SAR data provide greater details of the coastal environments and configuration of the shoreline and its changes. Extensive research has also been done using different interpolation techniques leading to the optimum convergence of the solutions. Thus, the basic causes, impacts and solutions regarding the natural hazard Coastal Erosion have been briefly discussed.

REFERENCES

1. Callaghan, D.P. and Wainwright, D. (2013) The Impact of Various Methods of Wave Transfers from Deep Water to Nearshore When Determining Extreme Beach Erosion. Coastal Engineering Journal, 74, 50-58.http://dx.doi.org/10.1016/j.coastaleng.2012.12.001
2. Kumaravel, S., Ramkumar, T., Gurunanam, B., Suresh, M. and Dharanirajan, K. (2013) An Application of Remote Sensing and GIS Based Shoreline Change Studies—A Case Study in the Cuddalore District, East Coast of Tamilnadu, South India. International Journal of Innovative Technology and Exploring Engineering (IJITEE), 2, 211-215.
3. Vinayaraj, P., Johnson, G., Udhaba Dora, G., Sajiv Philip, C., Sanil Kumar, V. and Gowthaman, R. (2011) Quantitative Estimation of Coastal Changes along Selected Locations of Karnataka, India: A GIS and Remote Sensing Approach. International Journal of Geosciences, 2, 385-393. http://dx.doi.org/10.4236/ijg.2011.24041
4. Davidson, M.A., Splinter, K.D. and Turner, I.L. (2013) A Simple Equilibrium Model for Predicting Shoreline Change. Coastal Engineering Journal, 73, 191-202.
5. Le Cozannet, G., Garein, M., Bulteau, T., Mirgon, C., Yates, M.L., Mendez, M., Baills, A., Idier, D. and Oliveros, C. (2013) An AHP Derived Method for Mapping the Physical Vulnerability of Coastal Areas at Regional Scales. Natural Hazards Earth Systems, 13, 1209-1227. http://dx.doi.org/10.5194/nhess-13-1209-2013
6. Dalrymple, R.A. and Rogers, B.D. (2006) Numerical Modelling of Water Waves with the SPH Method. Coastal Engineering Journal, 53, 141-147.
7. Van der A, D.A., Ribberink, J.S., van der Werf, J.J., Donoghue, T.O., Buijsrogge, R.H. and Kranenburg, W.M. (2013) Practical Sand Transport Formula for Non-Breaking Waves & Currents. Coastal Engineering, 76, 26-42.http://dx.doi.org/10.1016/j.coastaleng.2013.01.007
8. Muthukumarasamy, R., Mukesh, M.V., Tamilselvi, M., Singarasubramanian, S.R., Chandrasekaran, A. and Sabeen, H.M. (2013) Shoreline Changes Using Remote Sensing and GIS Environment: A Case Study of Valinokkam to Thoothukudi Area, Tamil Nadu, India. International Journal of Innovative Technology and Exploring Engineering (IJITEE), 2, 72-75.
9. Bouchahma, M., Yan, W. and Ouessar, M. (2012) Island Coastline Change Detection Based on Imaage Processing and Remote Sensing. Canadian Center of Science and Education, 5, 27-36.
10. Souza Filho, P.W.M. and Paradella, W.R. (2003) Use of Synthetic Aperture Radar for Recognition of Coastal Geomorphological Features, Land Use Assessment and Shoreline Changes in Braganca Coast, Param, North Brazil. Brazilian Academy Sciences, 75, 341-356.

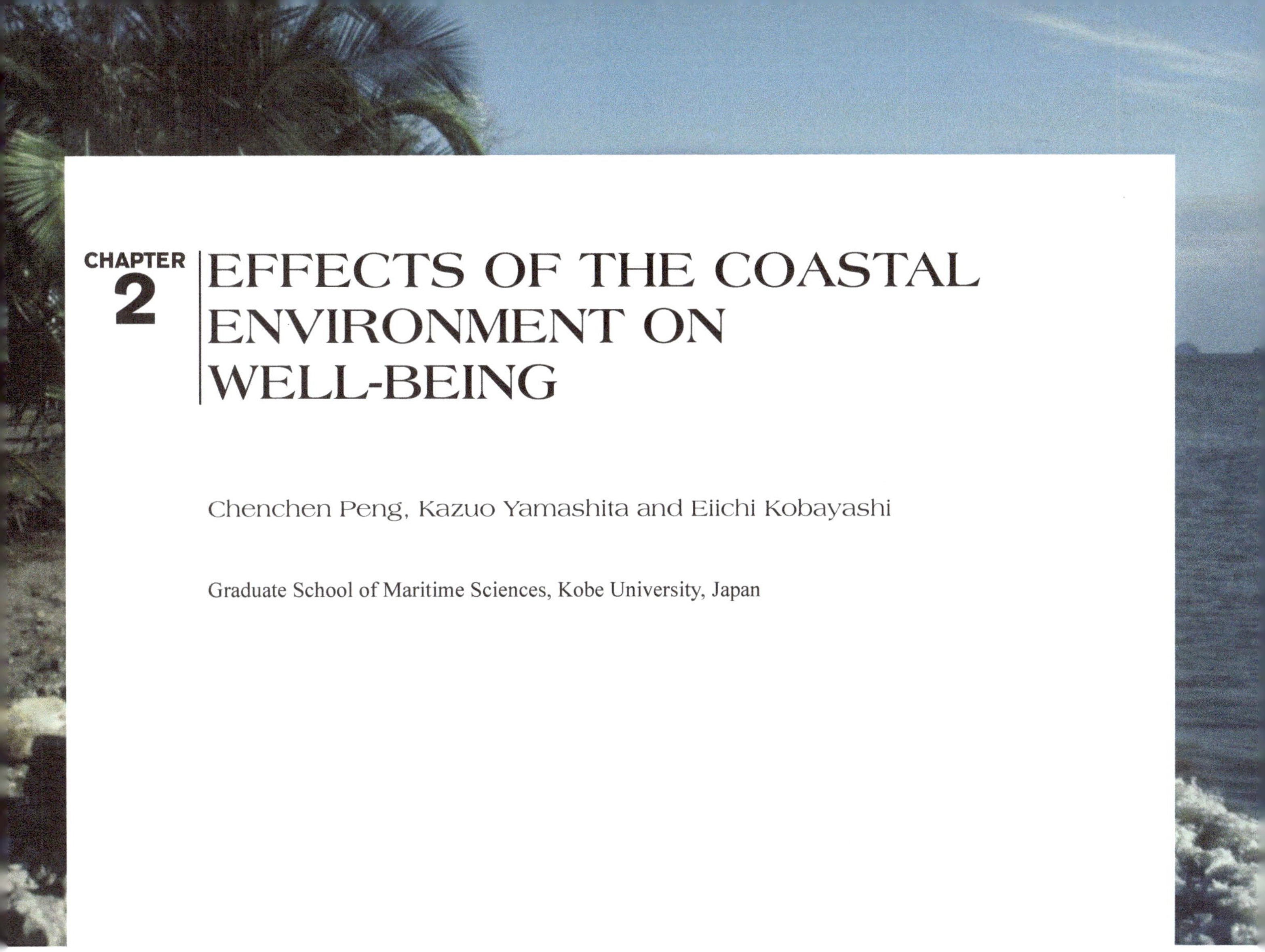

CHAPTER 2

EFFECTS OF THE COASTAL ENVIRONMENT ON WELL-BEING

Chenchen Peng, Kazuo Yamashita and Eiichi Kobayashi

Graduate School of Maritime Sciences, Kobe University, Japan

INTRODUCTION

Human beings and the ecosystem are always interrelated and interdependent. Humans change the natural environment they are dependent on. At the same time, the natural environment is also transforming humans. The conception of "cultural services" are focused on providing a framework for understanding human benefits from nature environment. Viewing nature fosters positive moods and improves mental health [1,2]. Kawakubo et al. [3] used fingertip plethysmography to confirm that contact with nature can benefit humans' health, such as by reducing stress. Some studies have demonstrated that time spent gazing at natural environments after a stress-inducing task leads to faster stress recovery than seeing urban environments [4]. Additionally, living in a natural environment is associated with fewer stress-related illnesses [5]. Evidence suggests that exposure to nature can have a range of psychological health benefits, including enhanced mood and mental wellbeing [6,7].

According to this theory, previous studies have also pointed out the positive impacts of aquatic settings on individuals' well-being, compared to those without exposure to such surroundings [8]. In particular, environments with water – positively regarded as natural "blue spaces" – are associated with higher preferences, greater positive effects and higher perceived restorativeness [9]. Tajima et al. [10] studied the effects of living near a waterfront area were both positive and high on people's daily lives. Generally speaking creating waterfront spaces in general, in cities such as with apartment complexes and housing, make people feel the "comfort of the waterfront area." Residents describe the coast as providing space, freedom, relaxation, and the ability to explore. In general, ocean views are believed to influence humans and result in positive psychological effects such as calmness and feelings of peace.

Over a third of the world's population naturally chooses to live along a "narrow fringe of **coastal land**" [11]. An analysis of panel data from England that explored the positive effects of coastal zones on improving health and well-being of people [12] showed that individuals reported significantly better general health as well as mental health when living closer to the sea [13]. Coastal environments have recently received growing attention as people have come to appreciate ocean views. One experiment showed the qualitative and quantitative relaxing effects of living by the sea, which were confirmed by visits to the beach [14]. Living in a coastal zone is an ideal lifestyle sought by residents.

Japan as an island nation that benefits from the sea that surrounds it on all sides. There are many people who

Citation: Peng C, Yamashita K, Kobayashi E (2016) Effects of the Coastal Environment on Well-being. J Coast Zone Manag 19:421. doi: 10.4172/2473-3350.1000421

live near the coast. We planned to use the self-administered questionnaire to investigate the different psychological effects on residents who live by the sea versus those who do not in Hyogo Prefecture, Japan. To choose this theme because many people recognized that the coastal **natural environment** play an important role in well-being. We hypothesized that having an ocean view from one's home would have an impact on residents' feelings; that it would increase positive psychological effects, and lower negative psychological repercussions. We investigated one coastal zone and one non-coastal area to find out that coastal environment altered individuals' well-being. This allowed us to gain insight into the connections between residents with ocean views and those without, based on gender and age. Coastal environments exerted significant influence on both males and females. In addition, the younger, middle-aged, and elderly groups living on the coast experienced more positive effects than those living in non-coastal areas.

METHODOLOGY

Subjects

We selected two residential housing areas in Hyogo Prefecture, Japan, one with ocean views and one without, to see how coastal zones influence well-being. Hyogo Prefecture has coastlines on the Seto Inland Sea and the Sea of Japan. All the residents selected in this study were no significant differences in the economic status.

We mailed questionnaire forms with a letter explaining the survey's purpose to randomly selected 4000 residents in both coastal and noncoastal areas. A total of 518 people answered the questionnaires, and returned the mailed questionnaires to us, for a collection rate of 12.95 percent. 518 questionnaires have well enough reliability and validity. Along with the questionnaire form, for ethical reasons we distributed a document describing the study's purpose, as well as a statement that participation was voluntary; this statement also mentioned privacyrelated issues such as anonymity. We assumed that people who returned the completed questionnaire in the prepaid, stamped envelopes had consented to participate. We conducted the survey from September to October of 2014.

Questionnaire

In this study, the self-administered questionnaire included questions about socio-demographics (gender and age), whether respondents had ocean views from their homes, and a **psychological effects** inventory.

In general, the question items were similar to the ones that Suzuki et al. [15] used in earlier studies. Suzuki et al. used the self-administered questionnaire to study the psychological effects of water area, however, we focused on the effects of coastal area on human well-being, so we modified the question items to facilitate our investigation of the effects of ocean views. The KJ method [16] is a quantitative and qualitative tool of building up a problem-solving method through repetition. We objectively analyzed the similar question items to the same subscales. We used the KJ method to classify 28 question items into five subscales: the passage of time (five items), magnitude and awe (six items), peace of mind (seven items), charm and longing (five items), and threat (five items). The names were selected for the five subscales (the passage of time, magnitude and awe, peace of mind, charm and longing, and threat) that clearly portrays the contents of the question items on our hypothesis. Four subscales of the questionnaire exert positive effects: the passage of time, magnitude and awe, peace of mind, and charm and longing. Threat leads to a negative effect.

Participants had to respond to each item on a 5-point Likert-type scale where 1 = "Strongly Disagree," 2 = "Disagree," 3 = "Neither Agree nor Disagree," 4 = "Agree," and 5 = "Strongly Agree." We scored each subscale by calculating the mean of the item responses.

Statistical analysis

A total of 518 residents who returned the questionnaires filled all the necessary information, so we calculated 518 questionnaire data in this study.

In our statistical analysis, we calculated the mean values of the inventory scores between residents with and without ocean views. Then, we compared the inventory scores between residents (with or without ocean views) based on gender. Thirdly, we categorized all residents into three groups according to age: younger (from teens to thirties), middleaged (from forties to fifties), and the elderly (from sixties to seventies). So the data could be compared inventory scores between residents (with or without ocean views) based on three groups.

We entered and analyzed inventory data using the software Statistical Package for Social Sciences (SPSS, 2007). In order to summarize and interpret the descriptive data, we conducted a one-way Analysis of Variance (ANOVA) and Tukey's HSD post-hoc analysis for multiple comparisons. We set the statistical significance level at $p < 0.05$, $p < 0.01$.

RESULTS

With/without ocean views

We received completed questionnaires from 518 participants; 301 (58%) had ocean views, while 217 (42%) did not.

Regarding the contrast between coastal and non-coastal environments, the data in **Table 1** show a significant difference ($p < 0.05$) for magnitude and awe, and strong significant differences ($p < 0.01$) for peace of mind, charm and longing, and threat. In terms of the passage of time ($p = 0.20$), there is no significant difference between the two areas.

Table 1: Predicting the psychological effects of residents with or without ocean views.

Subscales	Coast and Non-coast	
	F	*p*
Passage of time	1.60	0.20

Magnitude and awe	6.00	*
Peace of mind	19.50	**
Charm and longing	48.22	**
Threat	8.79	**

*p<0.05, **p<0.01)

Gender and with/without ocean views

The overall proportion of participants with ocean views consisted of 164 females (54%) and 137 males (46%); 128 females (59%) and 89 males (41%) did not have an ocean view.

In terms of the contrast between coastal and non-coastal environments based on gender, the data in **Table 2** show a significant difference ($p < 0.05$) for passage of time, and strong significant differences ($p < 0.01$) for magnitude and awe, peace of mind, and charm and longing on males. In terms of males, the threat ($p = 0.62$) is not significant between coastal and non-coastal environments. The impact of females significantly affects ($p < 0.05$) peace of mind, and has highly significant effects ($p < 0.01$) on the magnitude and awe, charm and longing, and threat. The passage of time ($p = 0.87$) is not significant.

Table 2: Associations between gender and with or without ocean views.

Subscales	**Coast/Non-coast and Gender**			
	Male		**Female**	
	F	*p*	**F**	*p*
Passage of time	5.08	*	0.02	0.87
Magnitude and awe	7.84	**	12.54	**
Peace of mind	20.88	**	4.16	*
Charm and longing	31.46	**	12.22	**
Threat	0.25	0.62	13.90	**

(*p<0.05, **p<0.01)

Age and with/without ocean views

In terms of the age distribution of participants with ocean views, 79 (26%) were in their 10s–30s, 110 (37%) were in their 40s–50s, and 112 (37%) were in their 60s–70s; regarding those without ocean views, 111 (51%) were in their 10s–30s, 77 (35%) were in their 40s–50s, and 29 (14%) were in their 60s-70s.

Regarding the differences between those with and without ocean views, and based on residents' ages, the results from **Table 3** show a significant effect ($p < 0.05$) on passage of time for the younger group. The magnitude and awe ($p = 0.57$), peace of mind ($p = 0.50$), charm and longing ($p = 0.20$), and threat ($p = 0.14$) are not significant. In terms of the middle-aged group show a significant increase ($p < 0.05$) on passage of time, and strong significant increases ($p < 0.01$) for magnitude and awe, peace of mind, as well as charm and longing. There is no significant on the threat ($p = 0.53$). In terms of the elderly group show significant increases ($p < 0.05$) on peace of mind and threat, and a strong significant increase ($p < 0.01$) on charm and longing. There is no significant on the passage of time ($p = 0.82$), and magnitude and awe ($p = 0.22$).

Table 3: Associations between three age groups and with or without ocean views.

Subscales	**Coast and Non-coast and Age**					
	Younger		**Middle**		**Elderly**	
	F	*p*	**F**	*p*	**F**	*p*
Passage of time	4.78	*	4.56	*	0.05	0.82
Magnitude and awe	0.33	0.57	7.89	**	1.51	0.22
Peace of mind	0.45	0.50	8.97	**	3.96	*

Charm and longing	1.63	0.20	25.01	**	16.59	**
Threat	2.27	0.14	0.39	0.53	5.31	*

(*p<0.05, **p<0.01)

Finally, in order to explore the relationship between the three age groups and the effects of coastal environments on well-being, we carried out a multiple comparison analysis, as shown in **Table 4**. The correlation between the younger and middle-aged groups is positive and significant ($p < 0.01$) for the passage of time, magnitude and awe, peace of mind, and charm and longing. The link between the younger and elderly groups is also strong significant ($p < 0.01$) for the four positive effects. However, the threat subscale shows no significant change in either the younger and middle-aged groups, or the younger and elderly groups. When comparing the middle-aged and elderly groups, there is no significant change on any of the five subscales.

Table 4: Turkey's HSD post-hoc analysis for three age groups living on the coast.

Subscales	**Coast and Non-coast and Age**								
	Younger/Middle			**Younger/Elderly**			**Middle/Elderly**		
	M1	**SE1**	*p*	**M2**	**SE2**	*p*	**M3**	**SE3**	*p*
Passage of time	-0.54	0.11	**	-0.58	0.11	**	-0.05	0.10	0.90
	-0.38	0.10	**	-0.33	0.10	**	0.05	0.09	0.86
Peace of mind	-0.32	0.10	**	-0.37	0.10	**	-0.06	0.09	0.80
Charm and longing	-0.42	0.12	**	-0.52	0.12	**	-0.11	0.11	0.60
Threat	0.12	0.16	0.71	0.19	0.16		0.06	0.14	0.89

*p<0.05, **p<0.01, M_1 =M(younger)-M(middle), M_2 =M(younger)- M(elderly), M_3=M(middle)-M(elderly)

DISCUSSION

After adjusting for gender, age, and with/without ocean views, there is an apparent increase in individual wellbeing when living in proximity to Japan's coast. The difference in the percentage of people reporting positive psychological effects in both coastal and non-coastal zones appears to strengthen with gender and age, but the difference is much stronger in **coastal areas**. The benefits of living near the coast may mitigate some negative psychological effects.

Coastal environments and psychological effects

In the results of this research (**Figure 1**), we can see the cause of a significant difference between coastal and non-coastal environments on the threat subscale. The coast may be important in reducing stress, and coastal zones may be considered therapeutic landscapes. This is important since visits to the coast are associated with especially strong feelings of restoration, and over time these feelings can help attenuate stress [17]. Compared with visits to non-coastal areas, the data indicate that visits to the coast are linked with higher levels of reduced stress and positive emotions. Watanabe et al. [18] supplied some supportive evidence on the restorative effects of exposure to the natural environment, which made viewers feel more relaxed. Engaging with nature heals the mind and body and promotes recovery from stress.

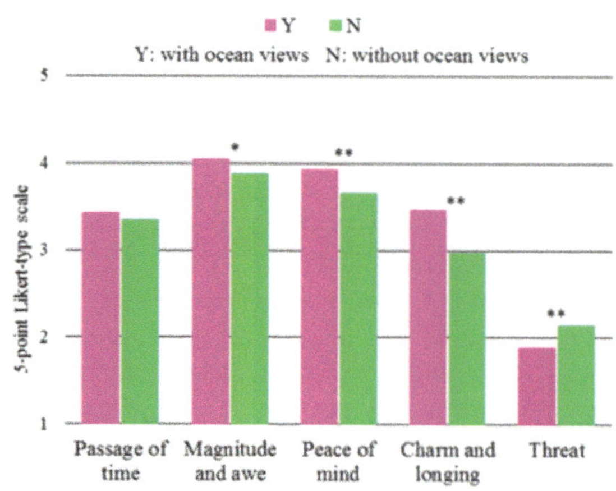

Figure 1: Mean scores for with/without ocean views (*p<0.05, **p<0.01).

Residents with ocean views emphasized the positive impacts they experienced from gazing at the sea, while those without ocean views did not express such feelings. The benefits of ocean views resonated with charm and longing, magnitude and awe, and peace of mind. People who live near the water believe ocean views are a good feature. Engaging with coastal environments may lead to better well-being due to the value of charm and longing that comes with greater leisure time spent near the sea. The closer people live to the ocean, the more likely they are to visit it. Access to the coast also seems to play a part in magnitude and awe as well. For example, walking on the beach is considered beneficial for positive emotions and feeling the power of nature. However, regarding the passage of time, there was no significant evidence that living near the ocean had any benefits compared to living away from it. We did not find any evidence of a beneficial effect of coastal proximity for the passage of time subscale.

In sum, coastal environments encourage a number of positive psychological benefits and behaviors that promote well-being; those who live near the coast are more likely to take advantage of these opportunities for increased wellbeing. The advantages of living near the sea are more strongly associated with increases in positive outcomes and reductions in negative ones.

Associating coastal environments and gender with psychological effects

The male-to-female ratio of the coastal and non-coastal environments in this survey was significant. Living near the coast was more strongly related to psychological effects for both men and women. In addition, the data revealed that women living near the sea had higher average scores than men for the four subscales (the passage of time, magnitude and awe, peace of mind, and charm and longing). On the other hand, they scored lower than men on threat.

Having an ocean view compared to those without an ocean views was significantly linked with positive impacts among males (**Figure 2**). Japanese men tend to have poor mental health and a higher level of work stress. It has been reported that men's poor mental health results from heavy occupational tension, and that lifecycles affect their mental health. Hence, men are a potential target population in the community for mental health support. We suggest that ocean views may be an instance of an important natural environment, which is a benefit for males to improve their mental health.

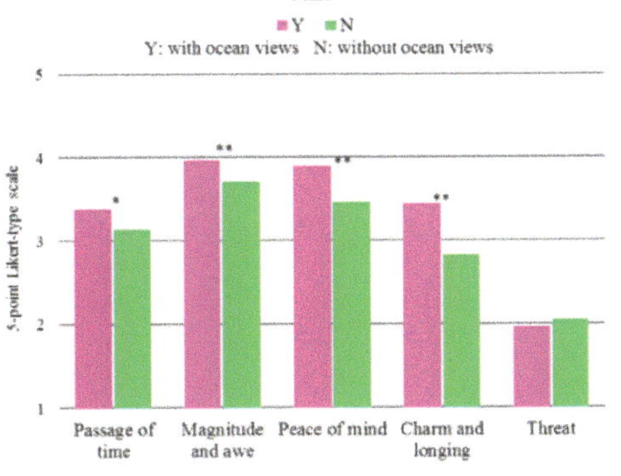

Figure 2: Mean scores for male and with/without ocean views (*p<0.05, **p<0.01).

We observed significant differences in positive effects among females between those with ocean views and those without (**Figure 3**). Above all, women living in coastal zones had a high numerical value; when questioned about gazing at the sea, they described feeling peaceful. The numerical worth of women who live with ocean views had a fairly high value for charm and longing. Regarding the passage of time, there was no difference in score for women with or without ocean views, but women with them had a higher range of error. Because Japanese women are busy to do housework, or take care of children every day, they could not look at time awareness, or feel that time is moving quickly or slowly.

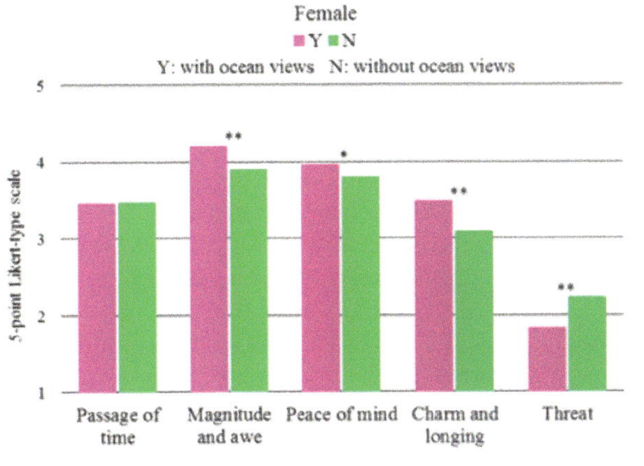

Figure 3: Mean scores for female and with/without ocean views (*p<0.05, **p<0.01).

As previously mentioned, women with ocean views showed more positive psychological effects as men. This may be due to traditional gender roles in Japan. There are many full-time housewives, and thus women have many more opportunities than men to look at the sea; staying at home and away from the office seems to have a positive correlation with health. Moreover, there was a significant difference in score only among women for negative effect of threat. The impact of threat on women living with ocean views was lower than for men. Because Japanese men are not only the cornerstone of society, but also the primary source of family. They have much more harmful impact than women. We presumed that the sea's freshness and openness influenced women more than men, and women were thus more able to resist the impact of negative feelings.

Associating coastal environments and age with psychological effects

We found that coastal effects were based on gender, but also on age. We could see the effect of age on coastal and non-coastal living, despite that we selected the participants from Hyogo Prefecture and that the sample size of the three age groups was not balanced.

We conducted each measure for the three age groups independently (**Figures 4-6**). There was a tendency for average values to rise as the age groups get older. The psychological impacts on the elderly group had higher average scores than the other groups. This difference could in part be explained by the residents' different backgrounds (e.g., age and sex) or by variations in the frequency and intensity of health-related issues. The link between those with ocean views versus those without in relation to individuals' psychological effects seemed to connect with age and gender. Individual well-being could be promoted by including more opportunities to view the sea. The impact of coastal environments on relieving stress and improving charm and longing were considered beneficial for well-being, more so than in non-coastal zones.

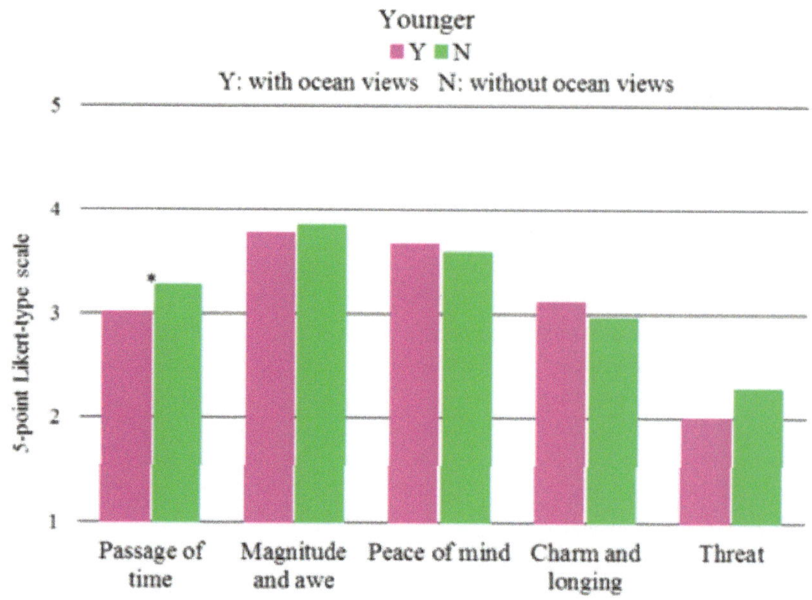

Figure 4: Mean scores for younger and with/without ocean views (*p<0.05, **p<0.01).

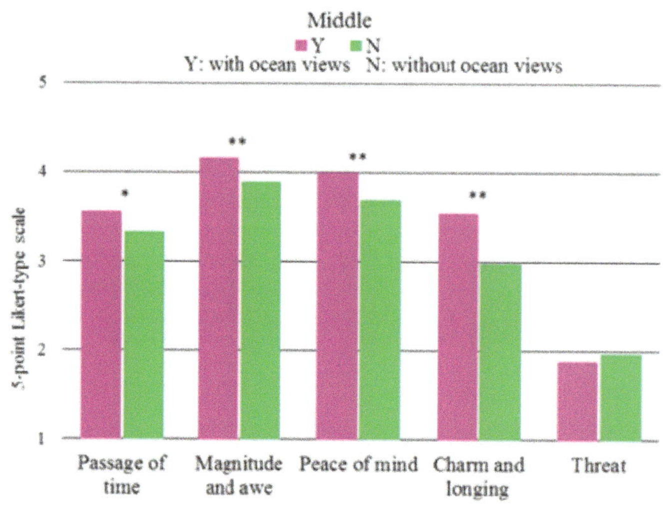

Figure 5: Mean scores for middle-aged and with/without ocean views (*p<0.05, **p<0.01).

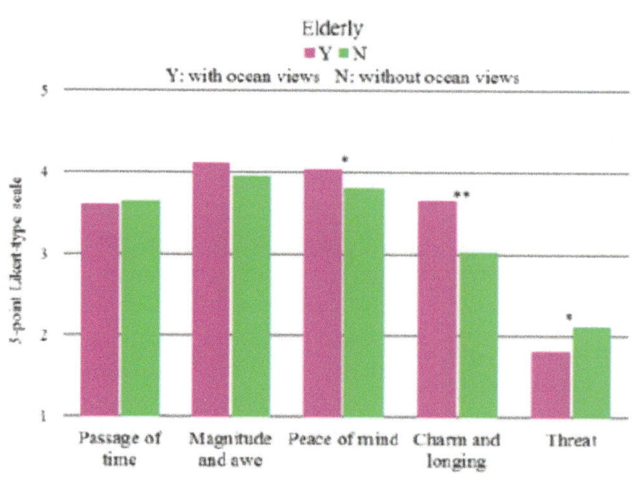

Figure 6: Mean scores for elderly and with/without ocean views (*p<0.05, **p<0.01).

With respect to psychological effects, we observed a significant difference between those with and without ocean views. The numerical, positive values of the younger group with ocean views were only somewhat higher than those without. This may be due to competition with schools and other social activities, so see less of an effect on younger people, whether or not they have an ocean view. However, the middleaged group living on the coast had a much, higher numerical value than middle-aged participants living inland. The middle-aged group had a strong sense of the effects resulting from ocean views in their day-today busy lives. Likewise, elderly people were much more likely to feel stronger psychological effects than those without. The elderly had more leisure time to enjoy the sea.

Despite living relatively close to the coast and perceiving health benefits, younger residents did not regularly enjoy ocean views or visit the beach. They often have little desire to appreciate the sea or go to the beach to play or relax. We assumed that all people would have a positive opinion of the sea. For example, on the beach, there is no need to look at or worry about the time. However, young people always have their mobile phones with them, which take up lots of time. Nowadays, even though young people spend less time in nature, merely living near the sea offers long-term exposure to ocean views, which is greater than for individuals without them. In terms of the passage of time between with ocean views and without ocean views, the average scores of older people were higher than the younger people, therefore, ocean views benefit older people and arouse fond memories of the passage of time more than younger people.

The middle-aged respondents with ocean views felt the sea's openness and vastness more positively than the younger group, despite their busy schedules. Our research showed that in terms of magnitude and awe, peace of mind, and charm and longing, they strongly appreciated and enjoyed the seaside. Among the positive effects we saw in the distinct scores among the age groups, the different average values of the middle-aged people between coastal environments and noncoastal environments increased in numerical worth more than those of young or elderly people.

Across all subscales, the factor of age demonstrated some significant positive differences. Compared to the other two groups, elderly people with ocean views were more likely to feel the passage of time, magnitude and awe, peace of mind, and charm and longing. We especially observed differences in the passage of time, as shown in the higher average value of the scores. It is said that as we get older, we have more opportunities to look back on the past. According to Yamazaki et al. [19], the frequency of reminiscence was shown to be higher among the elderly and middle-aged. We speculate that this is one reason why the elderly group scored high for the passage of time, since they enjoy opportunities to remember the past. Having time to remember an event may affect the frequency of the brain's ability to recollect. The elderly have a high tendency to reminisce, and time increases the frequency of recollection. In other words, the elderly have a positive, subjective sense of well-being associated with ocean views, based on a relaxed state that allows them to remember the past fondly. They have more time to stay at home, and thus find it easier to feel the charm and longing of the ocean than younger and middle-aged citizens. Young people have less interest in being in nature; this might be why they scored lower than the elderly on magnitude and awe, charm and longing, and peace of mind aroused images of ocean views.

Ocean views are believed to give us feelings of comfort and peace of mind, as well as a sense of well-being. The relationship between coastal zones as a healthy, natural environment that promotes well-being remains true, despite the danger that sometimes comes with it. The well-established and important literature on positive effects holds true.

CONCLUSION

We conducted a questionnaire (**Table 5**) survey to clarify the impact that coastal environments have on individuals' wellbeing; 518 residents living in Hyogo Prefecture's coastal and non-coastal zones participated. After adjusting for gender and age, we found an apparent beneficial effect from living in proximity to the coast. We obtained the following results.

Table 5: Questionnaire survey.

Questionnaire	
Subscales	**Question Items**
The Passage of time	You can forget how time flies.
	You can get tired of seeing the sea.
	You can feel getting a sense of season.
	You can forget everything easily.
	You can recall the old days.

Magnitude and Awe	You can feel magnitude and richness of nature.
	You can feel the mystery and wonder.
	You can feel the greatness and grandeur.
	You can feel the magical sense of natural beauty.
	You can feel a symbol of your own region.
	You can feel a sense of release and vastness.
Peace of Mind	You can cleanse yourself mentally.
	You can feel peaceful.
	You can feel pleasant.
	You can feel a sense of peace of mind.
	You can bring peace of mind.
	You can feel the freshness.
	You can feel the tranquillity.
Charm and Longing	You can get along with others easily.
	You can feel like a dream.
	You can feel elegance and luxury.
	You can feel like the place of relief.
	You can feel longing and hope.
Threat	You can feel intimidated.
	You can feel uncomfortable.
	You can feel upset.
	You can feel depressed.
	You can feel fear.

Firstly, compared to those who live inland, people who live near the sea showed higher positive psychological effects for the passage of time, magnitude and awe, peace of mind, and charm and longing. In contrast, the negative psychological effect of threat is lower for coastal residents. Living by the ocean leads residents to feel positive emotions (such as calmness and peace) and promotes general well-being.

Secondly, coastal environments had a greater positive influence on men than non-coastal zones. In addition, women living near the ocean had an increased likelihood of feeling more positive emotions in homes with ocean views, versus those without such views. Despite that coastal environments have been shown to exert influence on both males and females, these findings reveal that the seaside has stronger positive effects on females than on males. On the other hand, the negative effects of ocean views on females were weaker than on males. Thirdly, the younger, middle-aged, and elderly groups living in the coastal zone experienced more positive improvements than those in non-coastal parts. The three groups tended to demonstrate significant differences regarding the presence of an ocean view. The younger group is more likely to feel less positive effects than the middle-aged or elderly groups. The elderly group showed less negative effects than the younger and middle-aged ones. The positive consequences were strongest for the elderly due to exposure to ocean views.

The natural environment is transforming humans. Human psychology and behavior are dependent, not only on current social stimulus, but also on characteristics of the environment. If human behavior causes the living ecosystem to fall out of balance, it will have a serious effect on people's psychology and behavior, and may even be life-threatening. However, the natural environment can also have positive influences on human well-being. In summary, the potential benefits of coastal environments positively affected people. We recognize that coastal zones may enhance positive emotions. However, several shortcomings limit this generalization. We did not assume all potentially relevant variables; therefore further study, in conjunction with more detailed potential benefits of living in coastal zones and the effects of overdevelopment, are needed. We would need a larger sample size to clarify and expand upon our study.

We should consider the value of leisurely visits to the seaside to promote public health and psychological well-being. Even if it is not feasible for everyone to live in a coastal zone, some characteristics of coastal environments could be promoted to elicit relevant aspects of good health. Further understanding the effects of living near the ocean could be instructive for future interventions that encourage more people to spend time by the sea, or people elsewhere could experience elements of coastal environments.

REFERENCES

1. Alcock I, White MP, Wheeler BW, Fleming LE, Depledge MH (2014) Longitudinal effects on mental health of moving to greener and less green urban areas. Environmental Science & Technology 48: 1247-1255.
2. Thompson CW, Roe J, Aspinall P, Mitchell R, Clow A, et al. (2012) More green space is linked to less stress in deprived communities: Evidence from salivary cortisol patterns. Landscape and Urban Planning 105: 221-229.
3. Kawakubo A, Kasahara R, Oguchi T (2015) Effects of natural environmental picture and sound on relieving stress. Rikkyo Psychological Research 57: 11-19.
4. Van den Berg AE, Koole SL, Van derWulp NY (2003) Environmental preference and restoration: How are they related? Journal of environmental psychology 23: 135-146.
5. Grahn P, Stigsdotter UA (2003) Landscape planning and stress. Urban forestry & urban greening 2: 1-18.
6. Barton J, Pretty J (2010) What is the best dose of nature and green exercise for improving mental health? A multi-study analysis. Environmental science & technology 44: 3947-3955.
7. Thompson Coon J, Boddy K, Stein K, Whear R, Barton J, et al. (2011) Does participating in physical activity in outdoor natural environments have a greater effect on physical and mental wellbeing than physical activity indoors? A systematic review. Environmental science & technology 45: 1761-1772.
8. Wherrett JR (2000) Creating landscape preference models using internet survey techniques. Landscape research 25: 79-96.
9. White M, Smith A, Humphryes K, Pahl S, Snelling D, et al. (2010) Blue space: The importance of water for preference, affect, and restorativeness ratings of natural and built scenes. Journal of Environmental Psychology 30: 482-493.
10. Tajima K, Watanabe H, Kuroyanagi A (1997) A study on effect of urban waterfront area in high density living environment. Journal of Architecture, Planning, and Environmental Engineering 4: 277-284.
11. UNEP (2006, 2007) United Nations Environment Program Annual Report.
12. Wheeler BW, White M, Stahl-Timmins W, Depledge MH (2012) Does living by the coast improve health and wellbeing? Health & Place 18: 1198-1201.
13. White MP, Alcock I, Wheeler BW, Depledge MH (2013) Coastal proximity, health and well-being: results from a longitudinal panel survey. Health & Place 23: 97-103.
14. Miyamoto T, Shigeeda M, Sugiyama S, Oshima A, Togashi M (2001) Measurement of Relaxing Effects from a Visit to the Beach. Journal of International Society of Life Information Science 19: 38-46.
15. Suzuki N, Nakazawa O, Takahashi E, Yokomura Y (1985) Water's psychological effects gathered from questionnaires in Japan. Architectural Institute of Japan 413-414.
16. Kawakita J (1977) The KJ method for problem solving. Paper presented at the Marginal Land Problems Seminar.
17. White MP, Pahl S, Ashbullby K, Herbert S, Depledge MH (2013) Feelings of restoration from recent nature visits. Journal of Environmental Psychology 35: 40-51.
18. Watanabe T, Kokubo H, Takazawa K, Kawano K (2008) Psychophysiological changes during exposure to natural and urban environments. Journal of International Society of Life Information Science 26: 106-111.
19. Yamazaki S, Inatani F, Nonaka M (2010) Comparison in tendency and frequency of reminiscence between the elderly and the middle-aged: relation between quality of reminiscence and psychological well-being. Kurume University Psychological Research 9: 57-61.

CHAPTER 3

GEOMORPHOLOGICAL COASTAL CLASSIFICATIONS AFTER NATURAL PROCESSES AND HUMAN DISTURBANCE

Alcantara-Carrio J[1,2,4], Fontán Bouzas A[3], Albarracin S[2], Correa ID[4], Montoya Montes I[1] and Mahiques M[1]

[1]Oceanographic Institute, University of Sao Paulo. Pca Oceanographic, 191, Butanta, SP, 05508 900, Brazil

[2]Institute of Environment Science & Marina, Catholic University of Valencia, 94 Guillem de Castro, 46001 Valencia, Spain

[3]CIMA/University of Algarve, Building 7, Campus Gambel Faro, Faro 8005-139, Portugal

[4]Area of Marine Sciences, University EAFIT, AA 3300, Medellín, Colombia

CLASSIFICATIONS AFTER NATURAL PROCESSES

In geomorphological studies, a wide range of coastal classifications based on natural processes are available. These classifications results from very different approaches but are not necessarily incompatible. Used criteria basically include the tectonic [1] and hydrodynamic controls, as well as the sedimentological response.

Hydrodynamics control includes classifications after wave parameters [2,3], spring tidal range [2,4] and a combination of both [4-6]. A major advance in this type of classifications was the inclusion of the term megatidal, previously applied in ecological studies [7,8] and later adopted in coastal geomorphology. Thus, Levoy et al. [9] defined as megatidal the coast with tidal ranges higher than 8 m, while other authors [10,11] agree to locate the limit of megatidal coasts at 6 m. Therefore, the following four types of coasts, differentiated after its tidal range, should be considered as standards for future researches: microtidal (0-2 m), mesotidal (2-4 m), macrotidal (4-6 m), and megatidal (>6 m). A ternary classification, considering river, wave, and tide influence supported a very useful analysis of siliciclastic sedimentary coasts [12- 14]. The combination of tectonic and hydrodynamic controls led to the proposition of coastal morphogenetic classifications [15,16], which are probably the most widely used classification schemes. Grain size [17] and sediment composition are also very useful to classify and describe the coastal sedimentary environments. Thus, three different types of coastal environments are defined after their dominant mineralogical composition: siliciclastic, carbonate and evaporitic sedimentary coasts, main sediment sources relating to continental inputs, biogenic supply from the marine ecosystems, or evaporitic coastal processes, respectively. Therefore, mineralogy allows identifying the source and sinking areas of the coastal system [18]. Microfossil content in sediments is also useful to classify the coastal environments [19,20]. More accurate coastal classifications of sediment provenance can be obtained combining grain size, mineralogy and microfossil content analyses [21]. On the other hand, coastal classification based on availability or shortage of sediments, can be also obtained by combination of grain size and mineralogical features [22], as well as from analysis of the thickness of the sedimentary deposit [23].

COASTAL CLASSIFICATION AFTER HUMAN-DISTURBANCE

Kelletat [24] introduced for the first time the term manmade coasts. However, due to the complexity and diversity of human activities, a coastal geomorphological classification based on

Citation: Alcantara-Carrio J, Fontán Bouzas A, Albarracin S, Correa ID, Montoya Montes I, et al. (2014) Geomorphological Coastal Classifications after Natural Processes and Human Disturbance. Oceanography 2:e108. doi: 10.4172/2332-2632.1000e108

the extent of human disturbance has not been yet achieved. In order to cover this lack, we propose the definition of three main types of coasts according to the nature and extension of human disturbance: Natural indirectly modified coasts, Rural or intermediate coasts, and Human-developed coasts.

The term Natural indirectly-modified coasts refers to natural coastal zones without a significant human presence, but influenced by the global change, i.e. pollution, sea-level rise and consequent coastal erosional trends, and/or by human-induced loss of biodiversity, among other indirect anthropic impacts. Its geomorphological evolution is still mainly controlled by natural processes, but human impacts are also noticeable in minor proportions. Unfortunately, we must assume that currently all natural coasts are subject to human influences.

The class Rural or intermediate coasts applies where the human presence is low, typically associated to traditional practices, mainly agriculture and fishing activities. Land occupation in these areas is low or very low, with an alongshore population density lower than 10%. Therefore, the natural geomorphological processes are still dominant, but localized areas with noticeable human impacts can be easily identified.

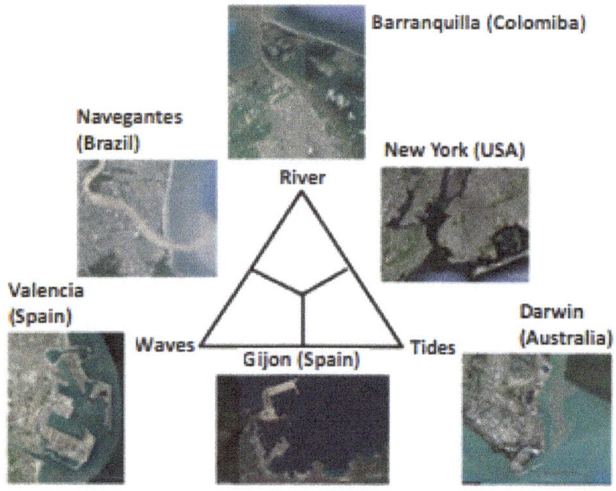

Figure 1: Harbors' classification after their morphology and coastal hydrodynamics (aerial photographs from Google Earth).

Human-developed coasts are those characterized by strong alterations of the natural processes, and natural landscapes are replaced by cultural landscapes (Nordstrom, 2000). This class of coasts includes Semi-urban coasts, Urban coasts, Manmade coasts, and Harbors. Semi-urban coasts are characterized by an intermediate alongshore density occupation, about 10-25%. They are typically occupied and developed by modern populations, with a higher number of socioeconomical activities and therefore, numerous direct impacts and substantial morphological modifications. For more human impacted coast, the term Urban coasts refers to those coasts occupied by cities, villages and settlements, with its shorelines covered in big proportion by walls, breakwaters, artificial reefs, promenades and other manmade constructions. Manmade or artificial coasts, sensu stricto, show strong modifications of its pre-existing natural morphology, or even events of total destruction and reconstruction, for instance in the artificial islands. Finally, harbors are another type of human developed coast, particularized because they are basically concrete coasts. Morphology of harbors results not only from engineering activities, but also from human adaptation to natural conditions. Thus, similarly to the subaerial morphological types of river deltas [25], they can be classified after the fluvial, wave and tidal hydrodynamic control (Figure 1). Fluvial dominated harbors locate on river channels, relatively close to the mouth. Tide dominated harbors are typical of macrotidal and megatidal estuaries, free of storm waves influence. Wave dominated harbors locate on open microtidal coasts without fluvial influence. Intermediate cases can also be defined, always with a logical relationship between the morphology and hydrodynamics.

Hard structures and beach nourishment are designed to protect intermediate and human-developed coasts from shore erosion, being isolated in the first case and very frequent in the second one. In contrast, alternative adaptation strategies like relocation and retreat are typically assumed in rural and low developed coast [26].

CONCLUSIONS

From a geomorphological point of view, the coast is a relatively narrow fringe whose morphology and evolution depend of numerous interactions between hydrodynamic, continental, atmospheric, tectonic, geochemical and biological processes, last one including the human disturbance.

After human disturbance, three kinds of coasts can be defined: Natural indirectly-modified coasts, Rural or intermediate coasts, and Human-developed coasts. Human developed coasts can be further subdivided in Semi-urban coasts, Urban coasts, Man-made coasts, and Harbors. The integration of previous tectonic, hydrodynamic, morphogenetic and sedimentary coastal classifications after natural processes, with this classification allows a holistic description of coastal geomorphology. Therefore, it should be useful to normalize the description of the study areas in coastal geomorphological researches, and consequently to facilitate comparisons among them.

REFERENCES

1. Inman DL, Nordstrom CE (1971) On the tectonic and morphologic classification of coasts. Journal of Geology 79: 1-21.
2. Davies JL (1964) A morphogenic approach to world shorelines. Zeitschrift für Geomorphologie 8: 1-42.
3. Walton TL, Adams WD (1976) Capacity of inlet outer bars to store sand. 15th Conference Proceeding,Coastal Eng, New York: ASCE.
4. Hayes MO (1979) Barrier island morphology as a function of wave and tide regime. In: Barrier islands from the Gulf of St. Lawrence to the Gulf of Mexico: Academic Press, New York, NY.
5. Nunmedal D, Fischer IA (1978) Process-response models for depositional shorelines: the German and the Georgia Bights.
6. Hayes MO (1975) Morphology of sand accumulations in estuaries: An introduction to the sympos.
7. Jouanneau JM, Latouche C (1982) Estimation of fluxes to the ocean from mega-tidal estuaries under moderate

climates and the problems they present. Developments in Hydrobiology 9: 23-29.

8. Koh CH (1997) Korean megatidal environments and tidal power projects: Korean tidal flats-biology, ecology and land uses by reclamations and other feasibilities. La Houille Blanche 3 : 66-78.

9. Levoy F, Anthony EJ, Monfort O, Larsonneur C (2000) The morphodynamics of megatidal beaches in Normandy, France. Marine Geology 171: 39-59.

10. Woodroffe CD (2003) Coasts: Form, Process and Evolution. Cambridge University Press.

11. Bird ECF (2003) Coastal Geomorphology, an introduction. John Willey and Sons, Chichester.

12. Boyd R, Dalrymple RW, Zaitlin BA (1992) Classification of clastic coastal depositional environments. Sedimentary Geology 80: 139-150.

13. Dalrymple RW, Zaitlin BA, Boyd R (1992) Estuarine facies models: conceptual basis and stratigraphic implications. J Sediment Petrol 62: 1130-1146.

14. Reading HG, Collinson JD (1996) Clastic coasts. (3rd edn). Sedimentary environments: processes, facies and Stratigraphy, Blackwell Science, Oxford.

15. Shepard FP (1976) Coastal classification and changing coastlines. Geoscience and Man 14: 53-64.

16. Shepard FP (1948) Submarine Geology. Harper & Row Publ, New York.

17. Friedman GM (1961) Distinction between dune, beach and river sands from their textural characteristics. Journal of Sedimentary Petrology 31: 514-520.

18. Calhoun RS, Fletcher CH, Harney JN (2002) A budget of marine and terrigenous sediments, Hanalei Bay, Kauai, Hawaiian Islands. Sedimentary Geology 150: 61-87.

19. Hippensteel SP, Martin RE (1999) Foraminifera as an indicator of overwash deposit, barrier island sediment supply, and barrier island evolution: Folly Island, South Carolina. Palaeogeography, Palaeoclimatology, Palaeoecology, 149: 115-125.

20. Leduc J Bilodeau G, de Vernal A, Mucci A (2002) Distribution of benthic foraminiferal populations in surface sediments of the Saguenay Fjord, before and after the 1996 flood. Palaeogeography, Palaeoclimatology, Palaeoecology 180: 207-223.

21. Alcantara-Carrio J, Fernandez-Bastero S, Alonso, I (2010) Source area determination of aeolian sediments at Jandia Isthmus (Fuerteventura, Canary Islands). Journal of Marine Systems 80: 219-234.

22. Alcantara-Carrio J, Alonso I (2001) Aeolian sediment availability in coastal areas defined from sedimentary parameters. Applications to a case study in Fuerteventura. Scientia Marina 65: 7-20.

23. Fontan Bouzas A, Alcantara-Carrio J, Montoya I, Barranco A, AlbarracinS et al. (2013) Distribution and thickness of sedimentary facies in the coastal dune, beach and nearshore sedimentary system at Maspalomas, Canary Islands. Geo-Marine Letters 33: 117-127.

24. Kelletat DH (1995) Atlas of coastal Geomorphology and zonality. Journal of Coastal Research, Special Issue 13: 286.

25. Galloway WE (1975) Process framework for describing the morphologic and stratigraphic evolution of deltaic depositional systems. Houston Geological Society, Houston.

26. Correa ID, Gonzalez JL (2000) Coastal erosion and village relocation: a Colombian case study. Ocean & Coastal Management 43: 51-64.

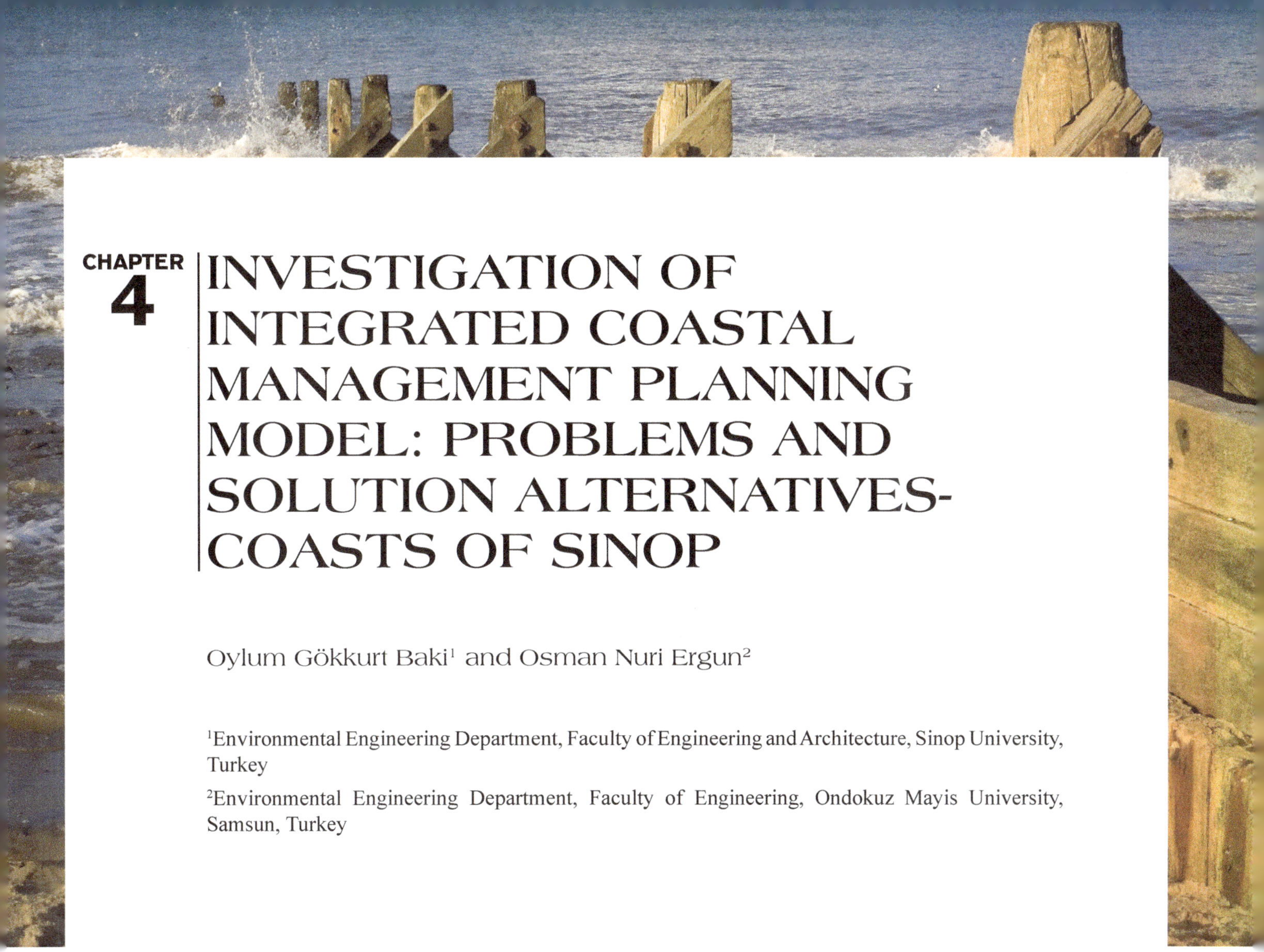

CHAPTER 4

INVESTIGATION OF INTEGRATED COASTAL MANAGEMENT PLANNING MODEL: PROBLEMS AND SOLUTION ALTERNATIVES-COASTS OF SINOP

Oylum Gökkurt Baki[1] and Osman Nuri Ergun[2]

[1]Environmental Engineering Department, Faculty of Engineering and Architecture, Sinop University, Turkey

[2]Environmental Engineering Department, Faculty of Engineering, Ondokuz Mayis University, Samsun, Turkey

INTRODUCTION

Urbanization and economic effects are causing new environmental problems and changes of current problems in terms of quantity and quality by putting pressure on coastal areas. This small area, where the two thirds of world population live [1], confronts us as the most vivid place where the social, economic and political relations of global system exist. According to [2], human pressure on the coastal areas causes deformation in natural structure and resources of the coast.

The units undergoing the fastest change under natural processes among various geographical formations of the earth are coastal areas. Being a center of attraction for many uses, coastal areas show the property of an important source for human life while confronting us as an important living area which needs to be protected in terms of natural wealth and biodiversity it contains due to being a transient area between land and coastal ecosystems [3,4]. Especially in the last 25-30 years, our coasts going under an intense population pressure based on the domestic and foreign tourism demand, building of second residences, excessive and unplanned housing, industrial institutions, use of coastal areas as disposal areas, uncontrolled sand extraction from beaches, coastal construction such as ports, piers, seawalls, etc. built without caring for city aesthetic and ecology, fillings made for land reclamations, excavations made in the name of coastal recreation and even highways constructed in the name of solution for transportation and many other practices clearly show how the coasts, which are the cradle of civilization and life lines of ecology, are under such a huge pressure.

There are two basic contradictions related to use of coastal resources. First; use of coastal resources to the benefit of humans and therefore necessity of consumption, second; necessity of preservation and protection of these resources for long term use. These two contradictions have become extremely critical stage in many countries, coastal areas and resources have non recoverably deteriorated [5]. Black Sea coasts, which have not lost its property of being an important coastal region in terms of commerce and culture from past to present, have its share of fast and unplanned urbanization in the recent years.

It is required to perform the management mentality in integrated mentality logic in the beginning of the coastal management, share the responsibilities with a multi-discipline working logic and plan the coordination as a structure which can be completely realized. Ensuring integration in terms of important integration matters of economics, socio-culture and environment is among the requirements of coastal management being sustainable [6]. In the process of coastal management, some measures may be taken to provide increase

Citation: Baki OG, Ergun ON (2016) Investigation of Integrated Coastal Management Planning Model: Problems and Solution Alternatives-Coasts of Sinop. J Coast Zone Manag 19:423. doi: 10.4172/2473-3350.1000423

in the effectiveness of practices and bring revenue. Taxing the polluters, taxation of manufacturing process, charges to collect from various economic activities or enterprises that may harm the environment or indirect taxes to be applied on use of motor vehicles and benefiting from coasts may be some of these measures. However, it can be said that measures such as these have not yet gotten a strong foothold in the process of coastal management and still in the development stage [7-9].

In this study, sustainable solutions and proposals were made and evaluations for the most suitable management model is conducted concerning coasts of Sinop province, which has become an important tourism locale in Black Sea Region in the recent years.

MATERIALS AND METHOD

Study area

Study area is Sinop city center which is at the northernmost point of Turkey and has the property of being a natural port, important with one of a kind structure on the peninsula at the coast of Black Sea (**Figure 1**). Sinop, at the Black Sea coast of the country, is a typical tombolo port.

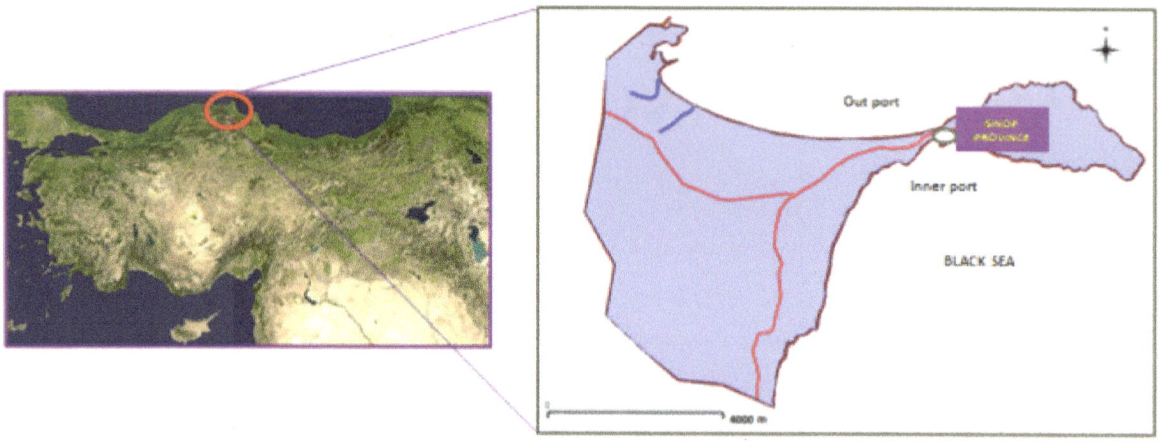

Figure 1: Study area.

Sinop is located at the Middle Black Sea Region. Established on Boztepe Cape and Peninsula found in the Black Sea coastline, extended towards north, the city is located between 41o 12', 42o 06' north latitudes and 34o 14', 35° 26' east longitudes.

Land border length of the province is 300 km, coastline length is 175 km. Has an area of 5862 km^2, which covers 0,8% of the country [10]. The peninsula of Sinop has a rising morphology from south to north. There are two ports in the city, one on the northwest and the other on the southeast.

Total population of Sinop is 1.984.412 according to 2007 census. 70th province (among 80 provinces) according to size of population, annual population increase rate of Sinop is 16.16% [11]. Between the years 1970 and 2000, urban population increase of Sinop is lower than Turkey urban population increase rate and shows a parallelism. However, population increase rate of Sinop city center in 1960-1965- 2007 was above Turkey urban population increase rate [14]. In general, number of people per km2 is 45 and in the city center, where the population is most sense, number of people per km^2 is 124.

Basic evaluations on current structure of coastal areas

The purpose in planning and protecting of coastal regions should be balanced development of coasts. Coastlines and coasts are important resources to satisfy the increasing recreation needs of town-dwellers [12,13]. In the coast management, the necessity of including the areas of the coast in two directions towards land and sea in management process as much as possible ensues [9]. For this reason, coasts are evaluated in a large scale as sea and land areas in this study.

The province of Sinop has 175 km long coastline on Black Sea. This coastline is usually straight. It extends in the northwest-southeast directions. There are cliffs in most of Sinop's coasts. In coastal areas where geomorphological conditions are convenient, ports can find the development environment.

In the study area, the biggest danger in terms of geology and geomorphology are landslides. In the land section of the study area, landslides are the most important of mass movements. Many landslides have happened in the region and still continues to happen based on the type and propery of rocks, incline of slope, precipitation and especially activities disturbing the natural balance such as cutting of trees, foundation excavation, road construction. Erosion based on wave activity throughout the coastline is high in the area. High and steep coasts seen in places where landslide areas connected with the sea are maninly in west and south of Sinop. Here, the coasts have highly degraded due to sea (wave) erosion and consists of sea cliffs 20-30 meters high [14-16].

System analysis in coast management planning used for province coasts

Three sub systems were defined in the study for the coast area. These three systems are in a physical interaction with each other according to scientific analysis. All human activities in close relationship with natural system is defined based on partly to physical behaviors of people, partly direct effects of use, partly physical infrastructure and partly to pollution due to wastes. These three main systems play a crucial role on coastal areas management.

In this study, three main sub-systems are defined for coastal management:

Natural system includes all interactions of all elements that are not human (atmosphere, lithosphere, hydrosphere) between each other via abiotic, biotic and chemical processes by their own dynamics.

User functions can be defined as all areas of interest of humans provided or able to be provided from natural resources.

Infrastructure includes all structures oriented to technical and organization. These infrastructures are necessary for the required user functions to exist and be used. In most circumstances, infrastructures cause planned or unplanned effects on natural system. These effects reach other user functions directly or indirectly and pressure and conflicts happen as a result (Adapted from [15]).

The purpose of coastal management is to understand, monitor and manage the processes between these three sub systems.

In the study, initially data were collected, areal planning and evaluation were conducted on coastal management of the province with the data obtained. Both verbal and written information and documents related to the subject were obtained from real and legal entities such as relevant public institutions like Provincial Municipality, Public Works and Settlement Provincial Directorate, Provincial Bank 16th Regional Directorate, the Provincial Directorate of State Hydraulic Works, Governorate Disaster Emergency Unit and Directorate of National Estate Revenue office, from private ordnance survey and law consultancy offices.

The gains to be obtained from the study are determined with the data collected. Coastal area system of the province is examined in all details, user functions are discussed as infrastructure (part of human system) and natural infrastructure system (part of nature system) in the system. By evaluating all data, a coastal management model of province was tried to be created. Besides that, institutions and organizations and objectives that may be effective in management planning.

The subjects for the SWOT analysis performed in the planning part of the study were determined, evaluations on these subjects were made. The purpose of the SWOT analysis is analysing the data obtained for the coastal management study of the province as best as possible and make a start to ensure the best practice of planning (**Table 1**). Accordingly the purpose of the analysis is to develop plans and strategies that will ultimately make use of strengths and opportunities, minimize the effects of threats and weaknesses by considering internal and external factors.

Table 1: Subjects Defined for SWOT Analysis to be conducted for Coastal Management Planning at the Province

Natural Structure	Economic Structure	Social Structure	Institutional Structure
Determination of natural structure properties	Determination of economic structure properties	Demographic and social structure	Determination of institutional structure properties
	Industry	Cultural and Natural Properties	
	Tourism	Residence	
	Transportation	Social Infrastructure	
	Logistics	Technical Infrastructure	
		Urban Living Quality	
		Urban Identity	

Through the problems revealed as a result of analysis, first the current problem was analysed and then probable solutions were set forth. During the management process, data obtained from analysis, institutional, legal and financial regulations were made use of and evaluated in terms of sustainable use of coastal areas.

RESULTS AND DISCUSSION

The city is a peninsula surrounded by the sea with a 175 km coastal line and a sustainable coastal management and suitable applications are a must to be supplied.In order to create a good planning a SWOT analysis was to be made so that coastal needs, advantages and disadvantages of the city could be seen clearly. The data gathered through SWOT analysis are demonstrated in **Table 1**.

Situation analysis (SWOT)

According to the prestudies and observations made on our study area and the experience coming from knowing the region, a Strengths, Weaknesses, Opportunities and Threats Analysis (SWOT) was conducted (**Tables 1 and 2**). In the analysis, Strengths and Weaknesses represent the internal data, in other words data on current situation of project area while Opportunities and Threats are external data.

Table 2: SWOT analysis results on coastal areas of the province (Updated from [17]).

SWOT MATRIX	**Strengths**	**Weaknesses**	**OBJECTIVES**
	Coastal region of Sinop being an attraction center at national level with its natural structure, morphology and undisturbed natural and cultural values in the back regions Sinop being a preferred tourism spot of Black Sea in the last 5 years at the national level **Coastal section not yet settled, creating opportunties for the future** Having critical and important areas in terms of natural life and having protection status concerning these areas **The presence of many geological formations like natural beauties such as coves in the coastal area** Nearly all settlements planning and implementary development plans being approved by Provincial Environmental Plan, municipalities having opinions and awareness on the competency status of these plans **Having a potential for yacht tourism and a promising future** Operation of Sinop Airport and connection of Sinop to all world via Istanbul, **Currently established Sinop University**	Summer tourism season being short due to climatic properties **Facility development at the coastal region being relatively low, tourism capacity being low** Tourism activities happening in the sea-sand-sun axis, tourism activity concentrating mostly on coastal region in Sinop province, **Yatch tourism and cruise tourism having negligible values in Sinop** Support uses and means (boat yard, boat maintenance and repair, etc.) to serve sea tourism being inadequate and undeveloped **Sinop Port having low passenger and freight capacity** Cooperation and coordination between institutions in planning and implementation at the coastal region of Sinop not being developed enough **Risks brought by economy of the city and the region considerably depending on agriculture and partially tourism, national and regional political processes and global economy changing tourism dynamism from year to year**	In this stage of the study, an objective matrix was established. Main objective determined for Sinop coastline is sustainable use of coastal area. This purpose can only be established in a context formed by environmental, economic and social balance. Determined objectives are as follows; 1. Taking environmental measures to increase sea water quality, 2. Protecting coastal areas against settlement, 3. Protecting historical structures at the coastal areas, restore and ensure to become the province protecting historical legacy, 6. Enabling use of beaches by developing and expanding, 7. Preventing and removing unplanned housing at coastal areas and behind them, 8. Determining sea pollution concentrations of land based polluters and turn sea environment back to its natural state.
	Opportunities	**Threats**	
	The concept of Integrated Coastal Areas Management being a worldwide approach and newly on the agenda and being an approach accepted internationally **Positive approaches of central and local administrations and non-government organizations** Important areas protected and undisturbed in terms of natural and cultural in the coastal region **Local administrations giving importance towards organization works of recreative use of coastal areas, especially in Sinop city center and districts,** Good will and accretion present in terms of tourism	Authority conflict between central institutions with planning and implementation authority and local administrations **Not being able to include the opinions of current University of the region on and about coast, deprivation of available knowledge** Having a need to re-organizing legal and administrative legislation for the best practice of country-wide coastal management **Many institution and organization having planning and implementing authority** High unearned income created based on development of tourism in the province dominating other types of use, other sectors not being able to compete with tourism	

Level of urbanization being low and industrialization being insufficient in Sinop province caused the result of environmental problems originating from urbanization and industrialization being at the lowest level, but still the natural, historic and cultural environment has started to partly deteriorate and get damaged. When compared to other parts of the country, it is known that pollution caused by urbanization and industrialization is lesser in the coasts of Sinop.

Management plans of sub systems

Current ports and shelters: With the data obtained as a result of interviews conducted with the fishing cooperatives making use of shelters based on official records of institutions responsible of coastal structurs of our country, current status of coastal fishing structures of Sinop province is revealed (**Table 3**).

Table 3: The fishing coastal structures and capacities in Sinop Province (modified from [18]).

Shelters	Capacity (boat and ship)
Akliman shelter and berth	200
Sinop Harbor	300
Demirciköy Harbor	construction has been completed in 2012

The fishing shelter found in the city center, due to having the property of being one of the most beautiful spots in the city, sometimes faces industry-specific and administrative problems. In parallel to increase in touring and private boats other than fishing boats in the recent years, it is observed that problems started to rise in matters such as anchorage and in-shelter organization by going over shelter usage capacity. Another coastal structure used in fishing activities is the natural harbor found in Akliman (**Figure 2**). The natural harbor at Akliman is a structure mainly used by fishermen, with a 111 m long pier, presenting risk of irregular berthing and sanding and without an infrastructure property other than the lighthouse at the pier [17,18].

Figure 2: (1st photo) Sinop Central Fishing Shelter (2nd photo) Sinop Akliman Harbor (Original, May, 2010) (3rdphoto) Demirciköy Fishing Shelter Auxiliary Pier Construction (Original, November, 2010).

Other than the fishing coastal structures, the port found in Sinop city center is open to passenger and freight shipment. Trade is made mostly to Russia and Ukraine from the Port of Sinop, which has import and export means. However, the volume of this trade is fairly low. Still, transit ships visiting the port, although rare, provides an economic gain. Between 1993-1998, citrus exportation was made to Ukraine. In 1995-1996, it was determined that exportation was made mostly to Ukraine in citrus [19].

Shelters belonging to fishing coastal structures are having overcrowding during tourism season, just like in fishing season. Coastal areas of the province with natural and historical wealth are used in service of tourism in certain periods of year. In summer, the number of amateur fishing boats and private yatches-spee boats are causing overcrowding in use of shelters due to location of shelter being close to city center and tourism potential. Also, yatches making use of shelter, while creating a good income source, in the progressing periods the shelter is wanted to be used for tourism purposes by leaving fishing activities out. The number of activities, which do not occur frequently in the province, should be increased. It can be said that this may happen by an improved management and tourism mentality.

Besides the ships coming to port in some years, Sinop may be expressed as a port city due to geographic properties, but the province is in a problematic location in terms of transportation. It is thought that if the operation of port is transferred to government from private sector, there would be improvements in the number or preference of passenger and freight ships. The port is in a notable position for the mobilization of tourism in the province.

Factors threatening coast and shore edge line

In Sinop and its surroundings, based on the rock type properties and incline of slopes, intense mass movements such as erosion or landslides are observed. Sea Wave Movements and the Effects of Landslide Structure on Formation of Coast is shown by the examination made on the regional geology map worked by MTA (Mineral Research and Exploration) as a factor to this situation, about 15-20 km of a distance from the shore has developed shallow and about 35-40 km from this point the shallowness turns into deepness of 200 meters then reaches about 1000 meters after that point [14]. This is important in terms of wave movements, which is one of the parameters affecting formation and determination of Shore Edge Line.

Waves, which the amplitude will grow by very severe wind movements, will not find a platform or a straight area to spread on like low-depressed shore formation, surging waves will be pressed into shore region and cause damage and erosion on the ground until their energy is broken and faded on the beveled cliffs from sea level. In the ground sections where ongoing active landslide is present, damage due to wave erosion is higher. It is possible to see changes due to natural causes on the shore edge line based on this change in shore morphology [20-23]. Current storey heights should be preserved in coastal areas with landslides. In coases with landslide structure and potential, residence height is lower than inner parts of the province. In landslide or potential landslide regions which are in development area, to ensure the residences to be built continue current pattern in a health way, it may be proposed to give construction factors as 3, at most 4-5 times. Storey additions or constructions, excavation and loadings should not be allowed in these areas. Also, garden irrigations should be controlled and measures should be taken to prevent water seepage to ground.

Ground movements can be seen in nearly all coasts of city center. For the purpose of preventing this, some solution proposals have been made. One of these solutions is the supporting walls seen in frequent intervals at the inner port beaches location (**Figure 3**).

Figure 3: Slope and Mobility Occurring Due to Landslides (Original, June, 2010).

Measures against slides seen more especially in the regions with second residences should be taken by relevant institutions. If the landslides occurring in the region cannot be prevented, it is thought there will high probability that plots in the coast area to be submersed. It is understood that the spread of sea waves on the parcels that are lost from the landslides will be an issue in time and stated that shore edge line is in the risk of going under the sea (**Figure 3**).

Current changes in shores

It was tried to determine the changes made or not in some places as a result of the land works conducted in the coastal area within this scope.

Non-intervened shores: As a result of land works, the places where interventions by human hands are made or not were determined and marked on the map (**Figure 4**). The land work conducted in the shores includes an area of approximately 35 km. When the area was analyzed, it was detected that 84% of the area were not exposed to any intervention (**Figure 4**).

Figure 4: Demonstration of intervened and non-intervened coasts of Sinop city center.

Intervened shores: As a result of the land works, the intervened areas which by some reasons were detected in the 35 km coastal area and they are marked in the map (**Figure 4**). Some changes were made in the shores of the 16% of the area. These changes can be listed as a changes resulted from the pressure on the grounds, buildings and roads, spurs and fill areas.

Fill areas

In 1981, there were lawsuits and registration lawsuits opened on the cancellation of 15 fill areas opened in the coastal area of Sinop city center and in order to demolish the buildings that were built against the law on the shore and coasts. These cases have been finalized.

12 cases that were opened again in the 1984 and finalized in the 1991-1992, related to prevention of intervention were gone to trial. Considering the contents, the cases were opened generally by people for registration application of the areas that are in the shore or coast line in their names and they were usually finalized as unfavorable.

The fill area on the road of Karakum showed in the **Figure 5** is one of the fill area that is in the legal process [24].

Figure 5: (a) old version of fill area on the road of Karakum. (b) New version of fill area on the road of Karakum (original, February, 2008) (c) Recreation area created in the inner harbor bay (Original, November, 2010).

The fill areas for recreation are available on the coastal area of Sinop City inner harbor (**Figure 5**). Detection of the use of coastal provinces of fill areas is important in terms of the implementation of Law No. 3621. At the same time, knowledge of the quality and the quantity of the coastal fill areas is a factor that can affect the features of the fill area. When the coastal area is considered as a shelter for the micro and macro level creatures, the importance of determining the fill areas becomes higher. The expansion works that were just started, which are approximately 106 meters long, are planned as a fill areas. It is thought by the authorities that the dock will expand 5 meters with this fill area and become suitable for the use of people. There will be a total of 7 fill areas in the coastal province including this one.

Changes of shore edge line

Determination of the situation of province's shore edge line and the screw plate of the shore edge line is a data necessary for the starting of coastal zone management.Determination of the coastal line as specified in the Coastal Law No. 3621 is also an important parameter, which specifies limits for the implementation of the law by identifying. The shore edge line enabling the usage of public and protecting the coasts, is stated as a first step in determining the usage rules for public welfare. 80 km of the 170 km coastal area in the province is determined and approved as the shore edge line. The areas which their shore edge lines are not identified must be done and included in the annual plan and completed by Governorship Shore Edge Line Identification Commission by taking those for settlement into consideration. The areas for identification of shore edge line without a 1/1000 or 1/5000 scaled base maps should be mapped by Provincial Special Administration in 1/1000 or 1/5000 scales. During the commission's works for identification of shore edge lines, adequate funding for making the drawings of the shore edge screw plates should be provided.

Under the Coastal Law numbered 3621/3830 and Zoning Law No. 3194 in order to prevent illegal housing made on the shore;

- Seminars and information meetings related to regulations of the Coastal Law and its application should be conducted for the village mukhtars whose villages are located in the coastal areas.
- Military police units for illegal structures should be established by Municipalities in the areas within jurisdiction of municipality, by Provincial Special Administration in the areas other than those and these units should continuously monitor and inspect whether illegal structures are built.
- The necessary interventions should be immediately made to the buildings which were started to be built illegally and completion should not be allowed. Also, illegally and unlawfully authorized buildings should be identified and processed according to Article 32 and 42 of Zoning Law No. 3194.
- According to the provisions of Section 5 of Unplanned Areas Zoning Ordinance issued according to Article 44 of Zoning Law No. 3194, structures built in villages especially on the shore edges should be built in accordance with engineering and health regulations and for the purpose of preventing illegal structures identification of village settlement areas and surroundings or when considered in larger scale, it will be appropriate that mapping within scope of Village Law No 442 should be performed within body of Provincial Special Administration or have them performed, in case of no funding regarding this issue, village grants from Provincial Bank Village Development Plan should be used for this purpose.
- Overall, in order to complete the identification of the shore edge line provincially and in Turkey general, the maps with 1/1000 or 1/5000 scale which deemed hard

to complete with local capacities should be provided as 1/5000 scale photogrammetrically topographical maps in coordination with the General Command of Mapping via mediation with the Ministry of Public Works, and shore edge line identification works would progress faster as a result of ensuring the development of undeveloped maps and delivery of these maps to provincial organizations,

- within scope of the Provisional Article 2 of Law No 2981/3290, for the purpose of resolving the buildings evaluated as against the registration and attachments, development of urban renovation projects by Mass Housing Administration in areas with dense settlement,
- Since the Zoning Law No. 3194 and No. 3621/3830 may not be considered separately, a revision must be made especially in the Zoning Law No. 3194 and ensuring integration with Law No. 3621/3830 within this revision,

In addition to this, in order to ensure the progress of the economy of the nation, providing special provisions for the Coastal Law No 3621/3830 within the scope of tourism areas, tourism centers and preventing construction of the structures in the coastal areas are opinioned [25].

Existing green areas in the province coastal areas

In the plains located in the coastal areas of the province, extending generally in the form of a narrow strip, settlement patterns show a typical residential property of the Black Sea Region. When the slope analysis of Sinop is examined, it is seen that usually 15-25% slope is present in the south- southeast of the peninsula. 0-40 % of the total slope change indicates that this area is appropriate for 2nd degree settlement [26].

Although Sinop is in the Black Sea region, green areas being fewer are remarkable (**Figures 6 and 7**). Green areas per capita in residential areas in the city are less than 7 m2/ person. In the developing area, the green areas are more than 7m2/person. compared to the standard green areas of 10 m^2 / capita, 30.5 hectares of green areas are needed, 3.8 ha of this are already present and 26.7 ha of green space is still needed.

Figure 6: (1st photo)a Greeen areas in residential areas on the expansion of the province

Figure 7: Diagram for Sinop coastal management plan proposal [27,28].

Accordingly, locations of the green areas of the city center and the coast are shown in the **Figure 6**. In order to prevent landslides and erosion, as mentioned above, the works or studies should be performed for the purpose of both supplying the settlement in the coastal line and fulfilling the needs of green lands. It is necessary to back up with a supporting wall where necessary and forestation of the specified areas in the city, avoiding multi-story buildings and overloading the ground. When the usage of area in the province's coast is examined, it was identified that the active green areas are very small and the available structuring is not in the homogeneous distribution. The walking, cycling and touring paths should be made available in the green areas and these are should not be restricted to guided tours. Temporary spaces should be made available to use as cafeterias or buffets that are least damaging to the environment.

Conclusion and Recommendations

Analyzing of the coasts of Sinop in terms of all economic, social and locational sectors were evaluated by considering the generally accepted Integrated Coastal Zone Management approach. In this context; it was determined how to use the coastal areas according to future sustainability principle and reasonably.

Provincial topics of sustainable practices are given in **Table 4**.

Table 4: Topics of sustainable practices and content for provincial villages.

Conservation of Nature and Natural Water Cycle	**Prevention of Improper Use of Natural Resources**
Providing sewage solutions for building constructions Not constructing new buildings without installing canalization systems and secondary improvement units and not using existing buildings Preventing all kinds of water pollution Creating separate canalization projects for rain water and used water Decreasing the negative effects of the recreational areas to environment – extending use of touristic areas Providing the alternatives to public by organizing recreational activities and decreasing the intensity at the beach areas in summer months	Establishing a unit responsible for the coordination between institutions Analyzing the usage of land in Sinop, in order to resolve the usage being contrary to each other, Determining the principles of usage and regulations for the Province Preparing the environmental impact assessment before the the new buildings project put into practice Issuing long term regulations for water and solid waste system Diversifying the touristic and economical activities in the region Diversifying the touristic activities including open-air activities Creating activity opportunities for winter tourism and the elder visitors.
Property of Being a Quality Tourism City	**Increasing Local Production Capacity and Improvement of Welfare**
Protecting the natural structure and landscape of the region Diversifying the tourism understanding as to include the environmental and cultural activities, Preventing the water pollution, Encouraging the people for environmentally friendly activities such as cycling, trekking and recycling, Providing information on environment to tourists and requirement of trained personnel, Providing sufficient information to domestic and foreign tourist guides on natural surroundings of Sinop their properties, Exhibiting traditional structure by bringing local foods, handworks and architectural styles.	Providing priority to local community in employment Facilitating entry of handworks, local foods and guidance to their industries Encouraging the touristic activities outside the hotel Providing tourism training including foreign language and environment information in occupational schools, Diversifying regional agricultural products according to requests of tourists Including environmental training and activities to curriculum of local public schools.
Increasing Local Production Capacity and Improvement of Welfare	**Minimizing the Negative Impact on Traditional Activities**
Providing priority to local community in employment Facilitating entry of handworks, local foods and guidance to their industries Encouraging the touristic activities outside the hotel Providing tourism training including foreign language and environment information in occupational schools, Diversifying regional agricultural products according to requests of tourists Including environmental training and activities to curriculum of local public schools.	Providing adequate recreational capacity to domestic tourists in forests and beaches, Stopping construction of summer houses and hotels until all legal, infrastructural and environmental issues are resolved.

The actors of commission required to be integrated and proposed to be formed according to sustainable use of provincial coast lines according to these studies are given in **Table 5**. Additions may be made which may be deemed necessary in implementation stage, other than those given in the list. Local administrations are evaluated on municipality basis and official regional and provincial administrations being present within this body will facilitate contacts with the administration. In scientific consultancy, relevant units of the university will provide solution and non-governmental institutions and other organizational structures are important in terms of participation and transparency, also in coordination in terms of public and coast lines. The diagram for Sinop coastal management plan proposal according to these studies are given in **Figure 7**.

Table 5: Sinop ICZM Planning and Institutions likely to be included in the application.

Local Administration	Regional and Provincial Directorate	University	Professional Chamber	Other Possible Partners
Sinop Governor (The one who is in charge of mak'ng decisions,approving thedecisions made , and leading the processes about everything related to the city)	**Public Works and Settlement Directorate** (The person who is responsible for determining the coast borders in the coastal area of the city and other housing planning, an important authority for coastal management)	**Sinop University (Instructors from Engineering Faculty-Environmental management department)** (It is highly important that coastal management studies should be supported by scientific data , which is significant for the sustainability and feasibility of the management. That is why, departments of universities that works on the subject should be involved in the process)	**Sinop Chamber of Commerce and Industry** (This chamber has to be involved in the decision-making process about any planning and application stage accompanied)	**Youth center** (The main aim to consider such an institution is that the opinions of young people will provide a good basis for the sustainability of the planning)
District Governorship (It is a subunit of governorship office and they perform as governorship offices in the districts)	**Provincial Directorate Of Culture and Tourism** (Since Sinop is a natural harbour and tourism center and a good coastal management planning and tourism should work together, this institution should be in the process)		**Chamber of Environmental and Civil Engineering** (They are two important occupation organization which are certified and authorized in the coastal management planning and application of the plans)	**Black Sea sailing club** (Since sailing is one of the most preferred activity of the coastal part of the city and the members of the union are sophisticated about the coastal area which makes them significant partners)
Sinop Municipality (It is the office which is responsible for environmental management systems which makes it a significant member of coastal management planning)	**Provincial Directorate of Environmental and Urbanization** (The opinions of this office on coastal management is utmost important in order to provide sustainability. Moreover, this office is able to consider coastal management with an environmental point of view)		**Union of Chamber of Merchants and Craftsmen** (The chamber should be authorized on the decision process of coastal area agricultural activities)	**Agriculture and stockbreeding development association** (During coastal area agricultural activities decision process, it is thought that this instutition can be utmost assitance)
District Municipalities (The office which exactly conducts what provincial municipalities do in cities)	**General Directorate of cultural and natural heritage** (After Sinop along with its coastal area has been a protected area on the second degree and this decision is inspected and applied by this instution)			**Aquacultural co-operative** (Fishing is almost the first source of income of the city people. Therefore, the opinions of Aqua Cultural Institution should be referred)
				Public participation (It is hard to succeed in any kind of planning or application without the consent and support of the public so the opinions of public should be asked)

Mixed measures should be considered in the province, for example, protection and development measures must be provided together. The results of the study, for the purpose of use and preservation of sustainable coastal ecosystems bring recommendations to the decision makers at national and local level, and it will be able to propose an institutional framework for implementation. Data obtained from the study will enable a status assessment for the study region and current status will be revealed and various scenarios will be created upon this status.

REFERENCES

1. Agenda 21 (1993) Programme of Action for Sustainable Development, Rio Decleration on Environment and Development Report of the United Nations Conference on Environment. United Nations, New York.
2. UNEP (1997) Water Branch-PAP Regional Activity Centre, Integrated Coastal Area and River Management. United Nations Environment Programme New York.
3. Erginoz MA (1998) Coastal using management discussion and recommendations. PhD thesis. Istanbul Technical University. Institue of marine Science and management. Istanbul: 203.
4. Sonmez R, Gorer N (1998) Policy and Strategies of Black Sea Integrated Management. Global environment fund. Ankara Turkey.
5. World Bank (1993) Guidelines for Integrated Coastal Zone Management. In: Post JC, Lundin CG (eds). Environmentally Sustainable Development Studies and Monographs Series 9: 16.
6. Kapdasli S (1992) Coastal engineering. Istanbul Technical University Publications. Istanbul: 277.
7. Vallega A (1996) Cityports, coastal zones and sustainable development. Cityports. Coastal Zones and Regional Change: International Perspectives on Planning and Management: 295-306.
8. Cicin-Sain B, Knecht RW (1988) Integrated Coastal and Ocean Management: Concepts and Practices. Island Press: 519.
9. Duru B (2003) Coastal policy, Integrated approches to coastal management and national coastal policy. Political Science graduate union foundation publications thesis series13: 379.
10. MEU (2009) Ministry of Environment and Urbanization, Sinop Kastamonu Çankiri Planning Zone 1/100.000 scale Environmental Plan Description Report. Environmental Impact Assessment and Planning General Directorate. Ankara Turkey.
11. MS (2009) Municipality of Sinop. Sinop-Erfelek Dam Water Treatment Plant and Transmission Line Construction Project Preliminary Report.Ministry of Environment and Forestry. DSI-State Hydraulic Works Water and Sewerage Department Ankara.
12. Duru B (2001) Integrated approches to coastal management and national coastal policy. PhD thesis. Ankara University Institue of Social Sciences: 145.
13. Kaya LG (2006) Critical Barriers to Rational Planning Processes for Coastal Zone Management: The Case Study of Antalya, Turkey. PhD. Dissertation. State University of New York, College of Environmental Science and Forestry, Faculty of Environmental Studies: 405.
14. MTA (1993) Land use potential Sinop. General Directorate of Mineral research and exploration Geological survey government Department. Ankara Turkey.
15. Coastlearn distance learning package for integrated coastal zone management.
16. IEPR (2009) 1/100.000 scale Istanbul environmental plan report. Section 4. SWOT analysis.
17. Baki GO, Ergun ON, Bat L (2015) Coastal management plan in the south of the Black Sea. Journal of Coastal Life Medicine 3: 600-606.
18. Baki B (2010) The current state of Sinop fishing coastal structures. Coastal and marine areas of Turkey national congress: 563-574.
19. MS (2005) Municipality of Sinop. Special Technical Specifications and Implementation Report Sinop-Erfelek Joint Solid Waste Disposal Facility.
20. MPWS (1985) The ministry of public works and settlement. Sinop Central District and Korucuk Village Geological Survey Report. General Directorate of Disaster Affairs Ankara.
21. MPWS (1988) The ministry of public works and settlement. Sinop Central District Geological survey report. General Directorate of Disaster Affairs Ankara.
22. MPWS (1991) The ministry of public works and settlement. Sinop Central District Geological survey report. General Directorate of Disaster Affairs Ankara.
23. MPWS (1993) The ministry of public works and settlement. Sinop Central District Geological survey report. General Directorate of Disaster Affairs Ankara.
24. Kosar G (2010b) Writen interview. General directorate of national real estate.
25. Inan T (2009) Interview. General directorate of national real estate.
26. MSU (2005) Study of Sinop zoning plan. Mimar Sinan University. Istanbul Turkey.
27. Baby S (2010) Techniques, Tools and Methods to Determine Coastal Morphologic Evolution and management for 'Coasts in Crisis' in The State of Kuwait. International Journal of Environmental Sciences 1: 9.
28. Proceedings of the Global Congress on ICM: Lessons Learned to Address New Challenges EMECS 10 - MEDCOAST 2013 Joint Conference.

CHAPTER 5

STUDYING HEAVY METALS IN SEDIMENTS LAYERS ALONG SELECTED SITES ON THE LEBANESE COAST

Nadine NASSIF, Ziad SAADE

Department of Environment, Agricultural Engineer, Lebanese University, Dekwanek, Lebanon

ABSTRACT

Ensuring the environmental protection of the Lebanese coast requires a continues monitoring system. For this purpose, four heavy metals (Fe, Mn, Cu and Pb in the marine sediments along the Lebanese coast were selected for analysis Sampling was carried out from two sites: Beirut and Batroun. Thus, 1g of dried sample is used for digestion by wet mineralization in order to determine the concentration of the four heavy metals by atomic absorption spectrometry. The results showed that Beirut area is polluted, by Fe and Mn as well as the station Bat 2 of Batroun. For Cu and Pb, Batroun region is more polluted in the superficial layers. The analysis also showed significant difference between the sites except for Cu. A difference between depths and between particles size fractions are observed for all the parameters studied. There is no a significant difference in layer sequence except for the Pb, and neither between the repetitions of the same sample. Results showed that the values of the four metals studied do not exceed the maximum limits at both sites, but they showed increase in comparison with the analyses obtained before July 2006 conflict, which was caused by the release of large quantity of fuel-oil from Jiyeh Power Station.

INTRODUCTION

The Lebanese coast has been threatened by a severe environmental disaster in its history due to the existed conflict in July 2006, which caused the release of 15,000 tones of fuel-oil after the destruction of oil tanks of the Jiyeh Electrical Power Station (30km south Beirut) [1]. This resulted a serious marine pollution, including a large portion of the marine ecology. This, in turn added a new aspect of pollution to the existing industrial, agricultural and urban ones, which cause complications on the aquatic environment that covers about 18,000 hectares (or 16% of the Lebanese territory), at a length of about 220km along the coast [2].

Among the contaminants brought about by human activities, heavy metals play a crucial role. Unlike most organic pollutants, heavy metals are natural constituents of the earth's crust, which are naturally released as a result of rock and soil weathering and erosion. The rapid population growth, is accompanied with several human activities, notably the industrial exploitation of natural resources, as well as sewage and refuse dumping into the sea without any treatment. They are often convoyed to the marine environment along rivers and perennial stream. Heavy metals are at low concentrations

Citation: N. NASSIF and Z. SAADE, "Studying Heavy Metals in Sediments Layers along Selected Sites on the Lebanese Coast," Journal of Water Resource and Protection, Vol. 2 No. 1, 2010, pp. 48-60. doi: 10.4236/jwarp.2010.21006.

in aquatic ecosystems where they tend to accumulate in marine sediments to reach toxic levels [3,4]. The difficulties in collecting valid samples for heavy metals analysis in marine water, made several researches focuses primarily on marine sediments where heavy metals are located at higher concentrations. Sediment is a compartment integrator of contaminants and can keep heavy metals among their layers and give an entire chronology of deposition mechanism.

Recent studies on marine sediments, conducted in Lebanon before the 2006 conflict, showed different heavy metal concentrations in different sites to the known concentration of heavy metals along the Mediterranean coasts [5,6].

In order to determine heavy metal content along the Lebanese coast, two representative sites were selected; Beirut (along the middle Lebanese coast) and Batroun to the north coastline. They are major ports, and then they contain a number of industrial activities. This study aims to evaluate the contents of four metals: lead, copper, iron and manganese. These elements are among the main elements released by industries effluents and different factories along the Lebanese coast.

The study aims also to carry out a comparative analysis of heavy metals at different sediments depth for each site, and even a comparison between the different depths in the same site that has not been addressed in the previous studies. The difference between repetitions made for the same sample (duplicate) was measured.

On the other hand, the results will help evaluating, for the first time in Lebanon, the influence of the size fraction to determine the most appropriate fraction containing the highest concentration of metals.

According to the concentration of heavy metals found in sediments in each region, we would be able to compare the degree of contamination before and after the 2006 conflict and then identify which of the two regions was more influenced by pollution.

METHODOLOGY

Sampling

The samples were selected from sediments cored from the two sites (**Figure 1**):

Site 1: Beirut (Ramlet el Bayda)

Site 2: Batroun (Selaata region)

In order to assess the detailed concentration of pollution, three stations were selected in each site. They diverse by the depth of sediment (**Table 1**).

Samples were taken by carrots, made of plastic material to avoid contamination, with 40cm length and 10cm diameter, these carrots were previously washed with HCL and HNO_3.

From each station, a spacing of 2 levies distant from each other by 30cm was done. The quantity of sediment samples was different from one place to another because of the granulometric so we had different amounts of sediment in the carrot (Figures 2 and 3).

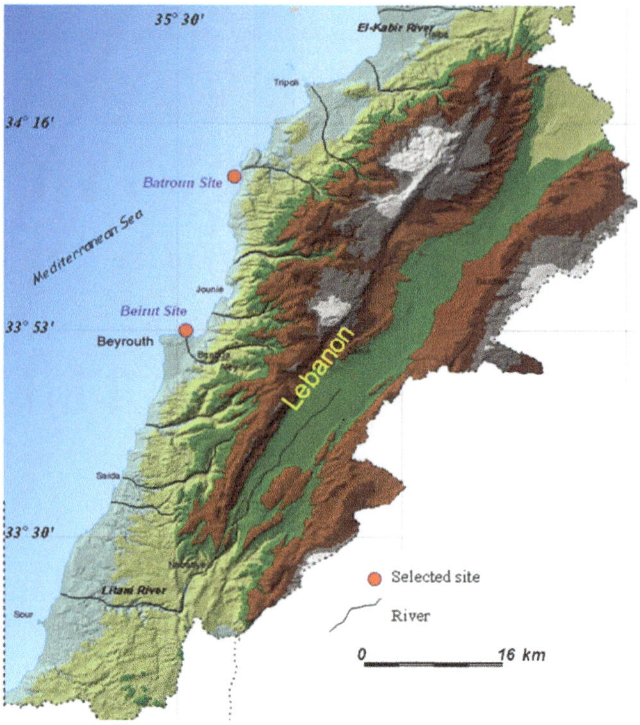

Figure 1. Localization of the selected samples.

Stations	Coordinates of the station	Depth
Beirut B 0m	N=33°52.550' E=035°28.820'	0m
Beirut B 5m	N=33°52.662' E=035°28.657'	5m
Beirut B 10m	N=33°52.688' E=035°28.458'	10m
Batroun BAT 1	N=34°15.092' E=035°39.211'	10m
Batroun BAT 2	N=34°16.169' E=035°39.226'	10m
Batroun BAT 3	N=34°16.709' E=035°39.115'	5m

Table 1. Location and depths of each station.

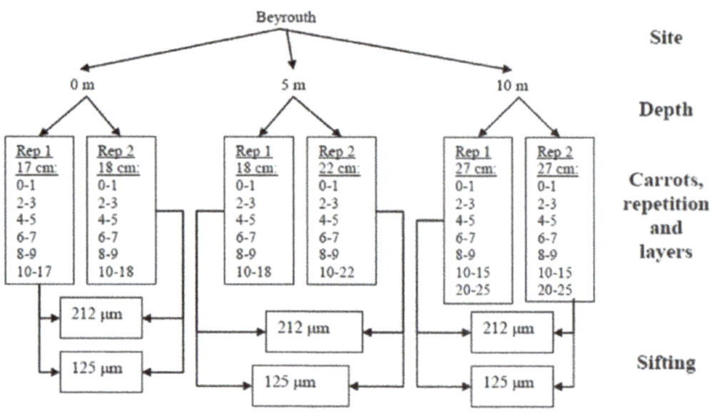

Figure 2. Schematic presentation of the sampling characteristics in Beirut.

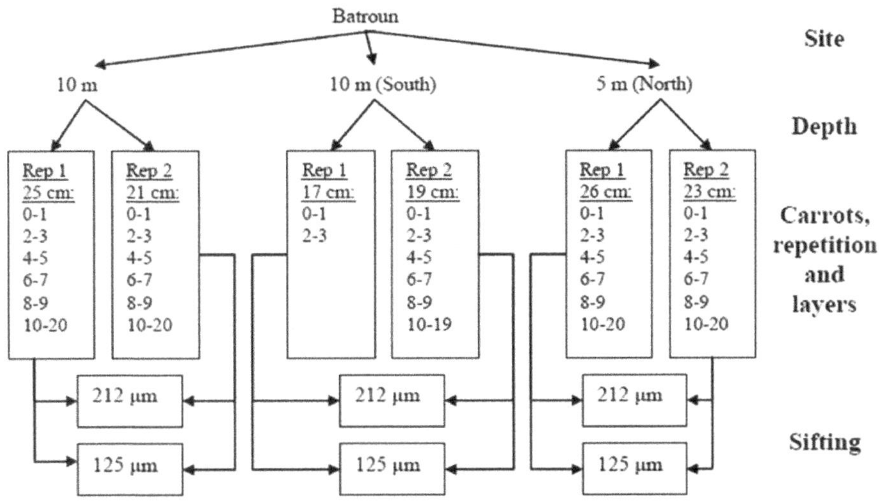

Figure 3. Schematic presentation of the sampling characteristics in Batroun.

Laboratory Analysis

1) Sampling equipments used in the field or the laboratory were maintained in safe condition and clean and had no trace of corrosion. They were handled and stored with the necessary precautions to avoid any aspect of contamination. All glassware used were previously soaked in 5% HCl, then 5% HNO_3 for 24 hours each, and then rinsed with double deionized water. All bottles used for transport and storage of samples were of polyethylene and cap perfectly sealed.
2) For the drying samples, we had cut carrots each 1 cm, and each 1cm of sediment was transferred and preserved separately in a plastic bag placed at room temperature for 8 days until we had a constant mass. This step, drying samples, is essential in order to improve the efficiency of mineralization acid, because the homogenization of dry samples is required other than wet samples. Once dried, the sediment is transferred separately in containers polypropylene previously labeled.
3) Grinding of sediments was carried out to increase the homogenization of samples.
4) Sifting was also applied to study the influence of particle size on the concentration of heavy metals. Sediments were sifted through sieves whose pores were respectively 212 µm, 125 µm and 38 µm [7].
5) Determining the size of the sample of sediments was the aim of this step to reduce the mass of an object without altering its other properties, and also to reduce the error to the least. The composition of the laboratory sample should be identical as possible to that of the initial sample [8]. For that, we had chosen randomly six samples from the three regions. For each sample, we crushed a small part of the sample, and then weighed four different mass: 0.25g, 0.5g, 1g and 2.5g. Then there will be 24 solutions which to mineralize, similar to the same way to determine their concentrations by Atomic Absorption Spectrometry (AAS).
6) Mineralization: is an essential step for the solution before analysis. To mineralize samples, the wet method was used, according to the following procedure, which was added to the test weighing, 20ml of HNO_3 (65%), on a hotplate, with magnetic agitation without boiling and evaporation. After six to seven hours, 5 ml of HCl (37%) were added every 30 minutes to three times until the total digestion. The acid then evaporated and the dry residue was recovered with 50ml of HNO_3, then we followed centrifuge and filtered in order to analyze them [9–11].
7) Analysis by Atomic Absorption Spectrometry allowed determining the levels of heavy metals by using a high temperature, the atomization, according to the case with a flame (for major elements like Fe and Mn) or an oven (for minor elements Cu and Pb). Consequently, all samples were measured with an Atomic Absorption Spectrometer (ZEEnit - 700) equipped with a system of correction Zeeman. A mixture of air-acetylene was used for (AAS) flame and argon for (AAS) oven. The curve of standard range with several known concentrations was primarily prepared. To analyze samples by (AAS), we need 1/7 dilution by HNO_3 (0.5N). Each sample was passed by a beam of light wavelength and intensity defined (λ_{Fe}=248.5 nm, λ_{Mn}=279.5 nm, λ_{Cu}=324.8 nm, λ_{Pb}=283.3nm) to measure their absorbance, then the concentration of each sample was identified, according to the curve of range standard of the metal [9,10,12].
8) After determining the size of sediment sample; however, all samples were mineralized, which were sifted. The mineralization was made for samples with the two fraction sizes 38-125 µm and 125-212 µm (since fraction <38µm is negligible). When the mineralization of theses samples has been done, the concentration was determined of every sample for each metal separately (for the two fraction sizes) by (AAS) with the same procedure followed in the determination of the size of sediment sample.

RESULTS

Five major findings were identified in this study, and then they were assessed to evaluate the degree of heavy metal contamination in the marine environment of Lebanon, as follows:

Percentage of Different Fraction Size in the Sediments Samples

Similar size fractions of different samples in each carrot were mixed to determine the percentages of each size fractions studied in each carrot of the two sites (**Table 2**). Results were plotted graphically (Figures 4 and 5).

Stations	Percentage of size fraction (%)		
	<38µm	*38-125 µm*	*125-212 µm*
B 0 m I	0.045	10.08	89.87
B 0m II	0.03	8.46	91.507
B 5m I	0.49	20.43	79.07
B 5 II	0.79	24.48	74.72
B 10m I	8.86	42.4	48.738
B 10m II	9.322	40.13	50.54
Bat 1I	0.176	4.44	95.35
Bat 1 II	0.082	3.39	96.48
Bat 2I	23.04	51.984	24.97
Bat 2 II	21.06	55.06	23.87
Bat 3 I	1.68	28.45	69.86
Bat 3 II	2.67	35.14	62.18

Table 2. Percentage of size fraction.

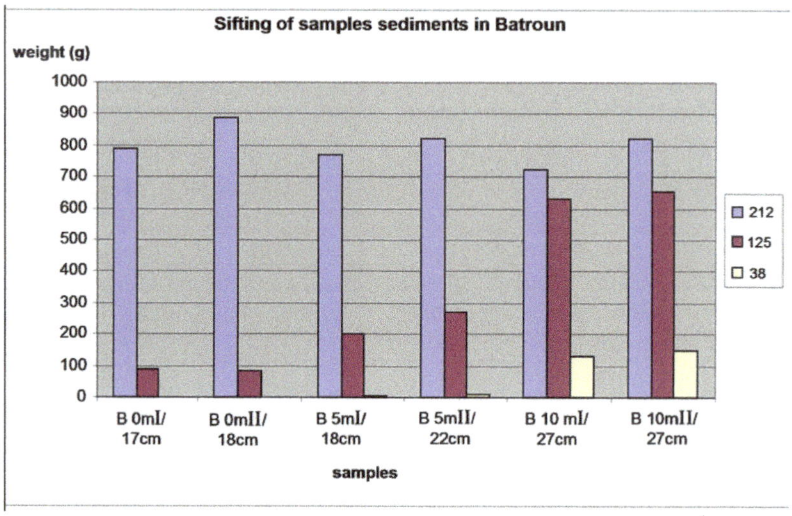

Figure 4. Distribution of the three fractions size in each carrot in Beirut site.

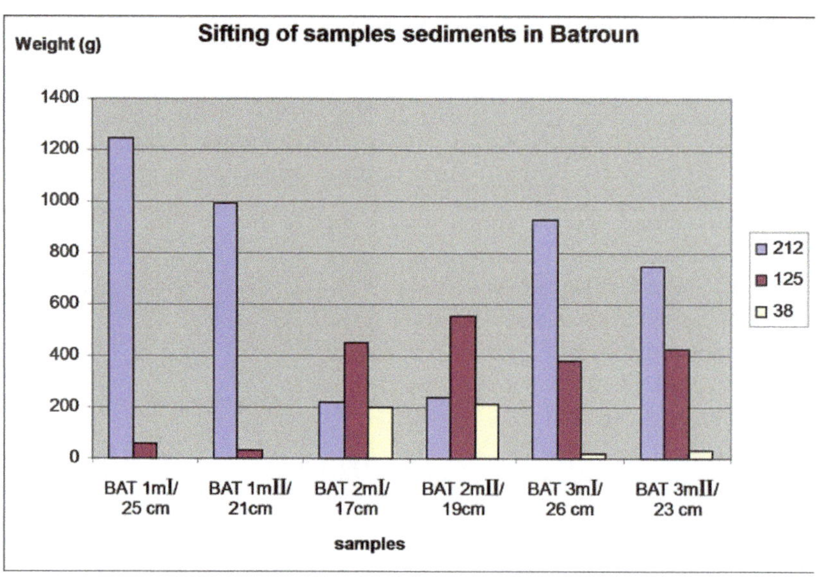

Figure 5. Distribution of the three fractions size in each carrot in Batroun site.

We note, similarly, that with increasing depth in Beirut site, the size fraction 125-212 μm was declined and the proportions of fractions <38 μm and 38-125 μm gradually increased. For the sample of Batroun site, the nature of rocks differs widely between depths, samples with the same depth Bat 1 and Bat 2 have a different composition. Comparing to the classification of sedimentary rocks in geology, the size fraction <38 μm is composed of silt and clay. The size fraction 38-125 μm was composed by fine sand and less silt, and the size fraction 125-212 μm was composed by fine sand and some sand.

Determination of the Size of Sediment Sample

The majority of concentrations on the three tests taken:

0.25g, 0.5g and 1g, treated in the same way, were proportional to the increase in weight of these three tests taken. However, this proportionality would no longer be applicable for most concentrations on the test 2.5g. The test sample of 2.5g requires a long period of mineralization and a high quantity of acid. In this study we used 1g as the size of sediment sample [13].

Concentrations of Heavy Metals

It was deduced that concentrations analyzed of Fe, Mn, Cu and Pb found in sediments of the two Lebanese sites (**Table 3**) were below the standards [14,15], and therefore they were acceptable, with the exception for some concentrations concerning the heavy metal Fe of the levies B 0m1 and the station Bat 2, which contain several values that transcend the limit and are polluted.

It is clear that in the two sites, samples from two levies of the same depth have the same proportions for each size fractions. The repeatability of samples was acceptable.

Analyses of Variance (SPSS)

The analyses of variance (SPSS) made for each site and for each heavy metal (Tables 4 and 5) showed a significant difference between the results for each studied setting.

B 5 m I. 212μm	[Fe] mg/l	[Mn] mg/l	[Cu] μg/l	[Pb] μg/l
0.25	51.89	1.645	38.199	16.086
0.5	100.7	2.445	55.104	33.208
1	223.7	3.755	101.85	58.903
2.5	541.4	4.133	207.9	89.46
B 5m II. 212μm				
0.25	48.17	1.669	19.533	22.568
0.5	92.04	2.353	34.125	28.37
1	183.5	3.305	73.01	49.595
2.5	518.4	4.2	93.03	97.37
B 5m I. 125μm				
0.25	159	2.041	78.23	9.201
0.5	338.6	2.971	120.61	16.814
1	654.4	4.93	251.88	32.207
2.5	1442	8.631	453.95	144.2
Bat 1 I. 212μm				
0.25	18.04	1.696	177.31	46.52
0.5	36.37	2.096	242.07	52.311
1	73.17	3.074	424.4	98.42
2.5	166.3	3.733	559.35	250.2
Bat 1 II. 212μm				
0.25	16.15	1.229	44.98	18.865
0.5	33.5	2.139	168.38	33.572
1	65.86	3.408	283.29	72.05
2.5	140.2	3.998	374.55	103.81

Table 3. Concentrations of Fe, Mn, Cu and Pb to six randomly sample for 4 different sizes of sediment sample.

Effect	Site	Depth	Layer	Fraction	Repetition
Fe Beirut	+	-	-	+	-
Fe Batroun	+	+	-	+	-
Mn Beirut	+	+	-	+	-
Mn Batroun	+	+	-	+	-
Cu Beirut	-	+	-	+	-
Cu Batroun	-	-	-	+	-
Pb Beirut	+	+	-	+	-
Pb Batroun	+	+	+	+	-

(+: Significant difference; -: no significant difference)

Table 4. The effects of different settings for the Fe, Mn, Cu and Pb.

Concentration of Fe

Compared with the concentration of Fe samples (with a diameter <200 μm) taken from the pit of Beirut in 2004 (41.87mg/l) [13], it was found that concentrations of Fe are largely higher due the impact from the conflict of July 2006.

Figure 6(a), (b) and (c) were studying the change in the concentration of Fe, each one concerned a specific layer, between the different sampling made in the two sites and for the two size fractions selected before. In these three layers, for both studied size fractions, the station Bat 2 site was the most concentrated in Fe. The three stations of Beirut had close concentrations of Fe, lower than Bat 2, but higher than those Bat 1 and Bat 3, which had very low concentration of Fe. In the site of Beirut and for the size fraction 125-212 μm, the concentration of Fe in station B 10m is slightly higher than in B 5m which was in turn slightly higher than B 0m. For the size fraction 38-125 μm, the concentrations of Fe in the three depths were very close to each other.

Figure 7 shows the change in the concentration of Fe according to the different layers of a levy. The various points 0, 1, 2, 3, 4 and 5 of the y-axis represent; respectively the first, third, fifth, seventh, ninth and eleventh layers.

The charts in **Figure 7** show that there was no significant difference between layers of the same depth, except for Bat 2, which shows some variation between layers because of the heavy concentration of Fe.

Almost all charts show that repetition did not make a significant difference for the concentration of Fe with the exception of Bat 2 for the same reason cited above. It was also observable that the samples with the finer size fraction (38-125 μm) are those who generally greater concentrations of Fe compared to the other fractions, this effect was also observable to the samples of Beirut.

Concentration of Mn

Figure 8(a), (b) and (c) shows the change in the concentration of Mn, each one concerned a specific layer, between the different sampling made in the two sites and for the two size fractions selected before. The results show the difference between the two sites. The station Bat 3 is the least concentrated among all others and for both size fractions. The two stations B 10m and Bat 2 are most concentrated for the two size fractions. In Beirut, B10m had the most concentrated in Mn, then B 5m and then B 0m.

The results for Mn show that samples of Beirut and Batroun did not differ significantly according to the layers.

It is obvious from Figures 8 and 9 that there was no significant difference for the repetition of the samples. Moreover, the results show that in the site of Beirut the size finer fractions contain a high concentration of Mn. While samples of Batroun did not show a difference between the two size fractions, with the exception of Bat 2, which contained a high concentration of Mn.

Concentration of Cu

Compared with the concentration of Cu samples (samples with a diameter < 200 μm) taken from the pit of Beirut in 2004 (33.67 μg/l) (13), hence, the concentration of Cu far exceeded the previous values after the conflict of July 2006. **Figure 10**-a, 10-b and 10-c reveals change in the concentration of Mn, each one concerned a specific layer, between the different sampling made in the two sites and for the two size fractions selected before. For the size fraction 125-212 μm, results show that both sites were highly concentrated and had almost the same concentrations. While, Bat 2 and B10m presented slightly higher concentrations than the other stations.

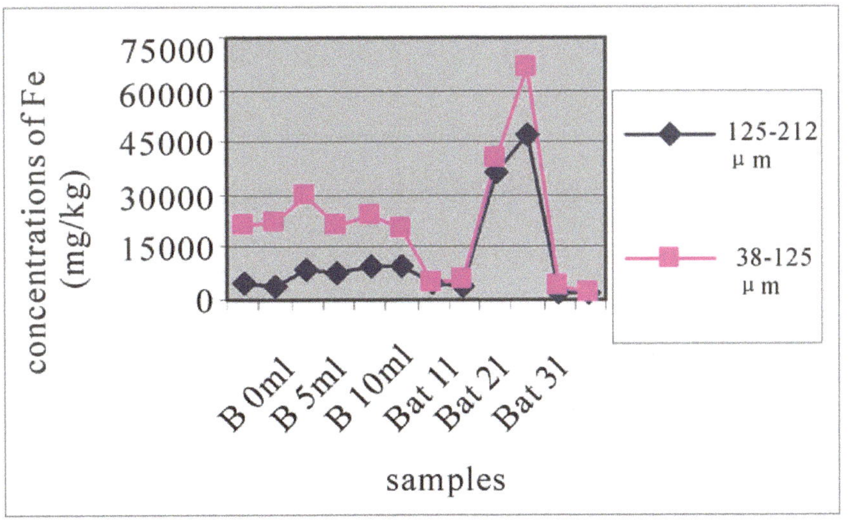

(a)

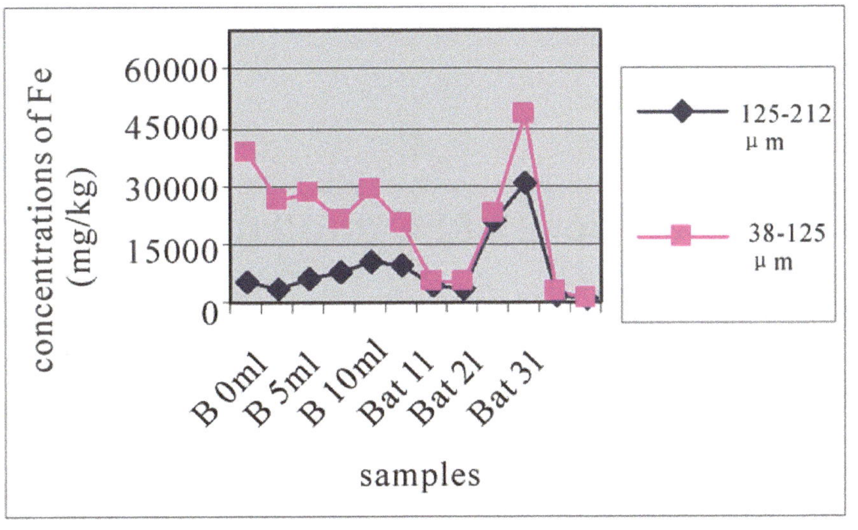

(b)

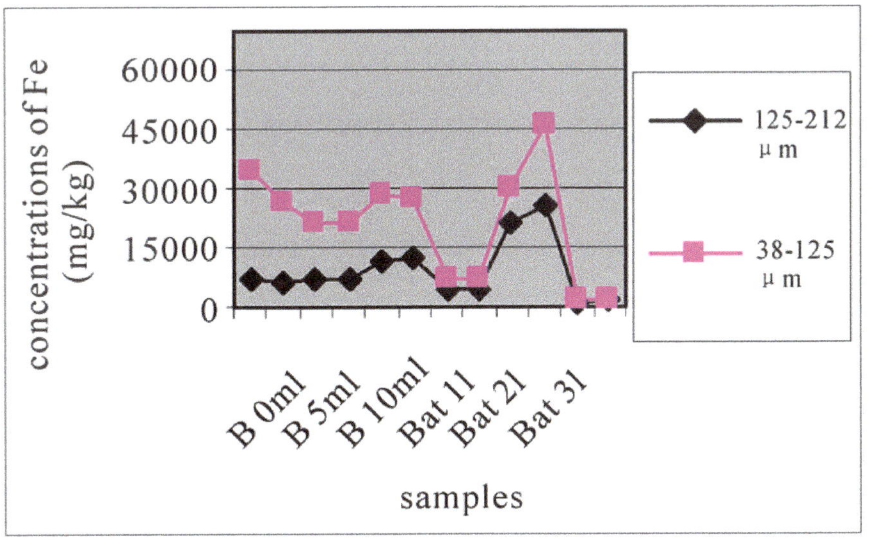

(c)

Figure 6. Change in the concentration of Fe for the both size fractions of the first layer (a), the seventh layer (b) and the eleventh layer (c), in the different levies of the two sites.

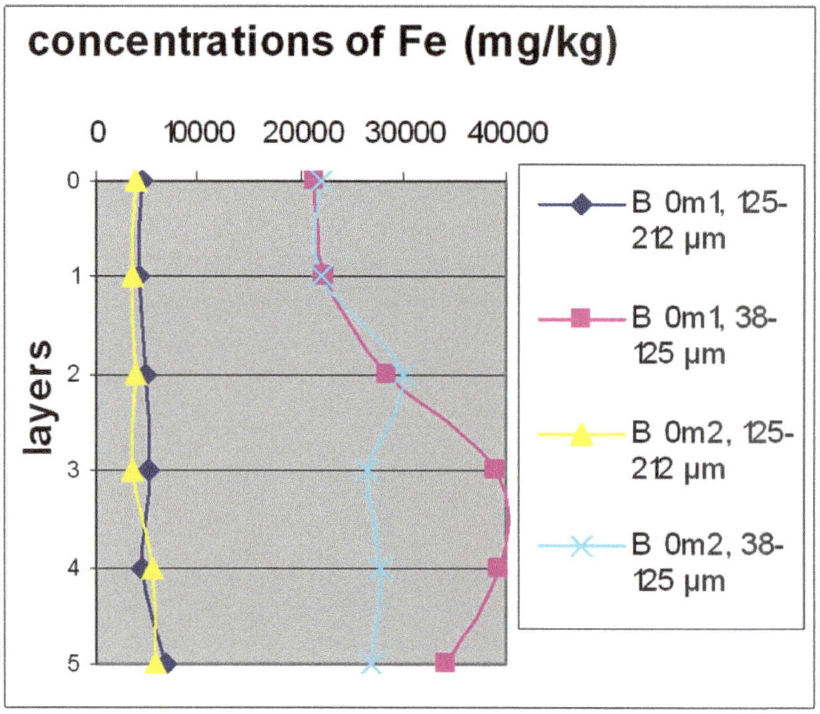
concentrations of Fe (mg/kg)
0
10000
20000
30000
40000
layers
0
1
2
3
4
5
B 0m1, 125-212 μm
B 0m1, 38-125 μm
B 0m2, 125-212 μm
B 0m2, 38-125 μm

(a)

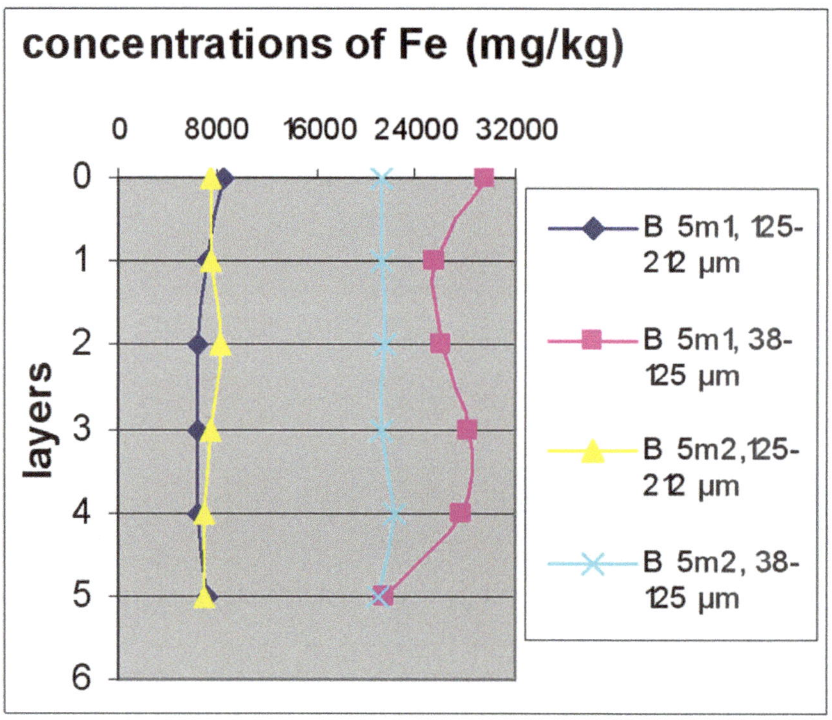
concentrations of Fe (mg/kg)
0
8000
16000
24000
32000
layers
0
1
2
3
4
5
6
B 5m1, 125-212 μm
B 5m1, 38-125 μm
B 5m2,125-212 μm
B 5m2, 38-125 μm

(b)

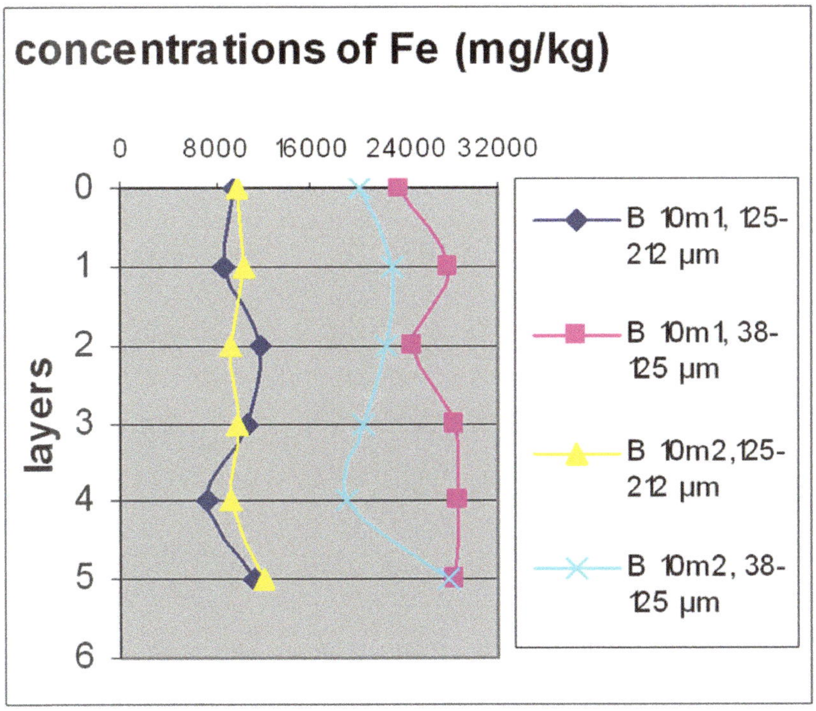
concentrations of Fe (mg/kg)
0 8000 16000 24000 32000
0 1 2 3 4 5 6
layers
B 10m1, 125-212 µm
B 10m1, 38-125 µm
B 10m2,125-212 µm
B 10m2, 38-125 µm

(c)

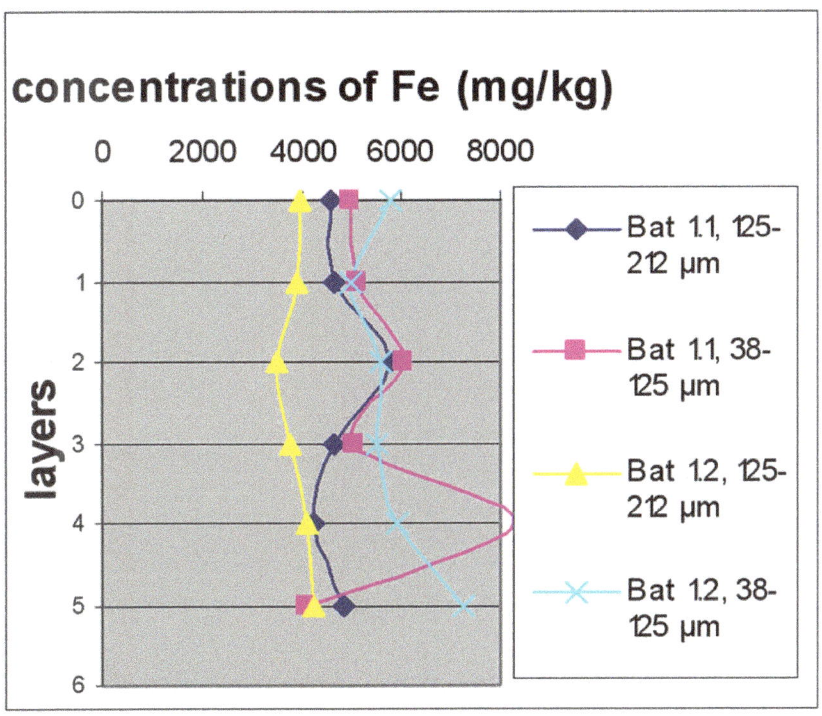
concentrations of Fe (mg/kg)
0 2000 4000 6000 8000
0 1 2 3 4 5 6
layers
Bat 1.1, 125-212 µm
Bat 1.1, 38-125 µm
Bat 1.2, 125-212 µm
Bat 1.2, 38-125 µm

(d)

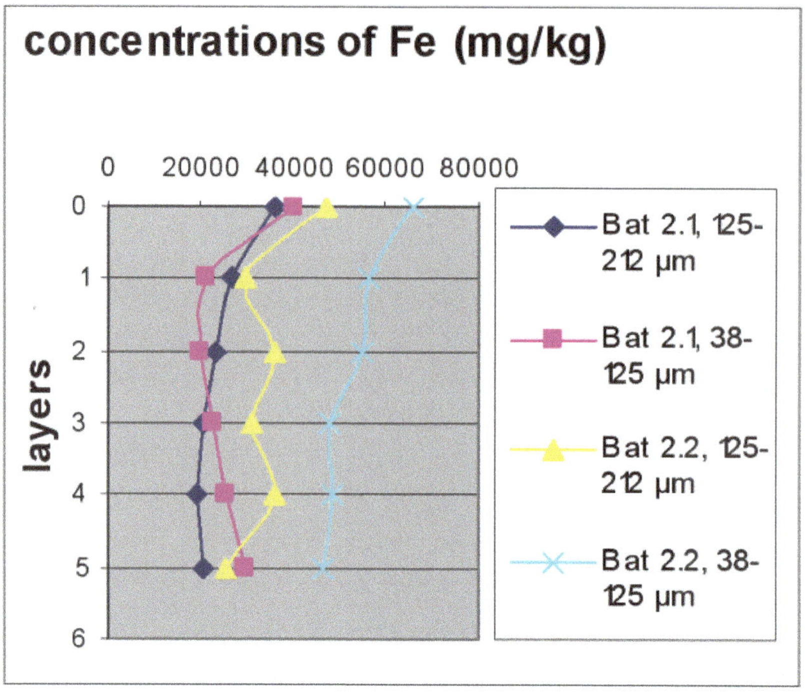

(e)

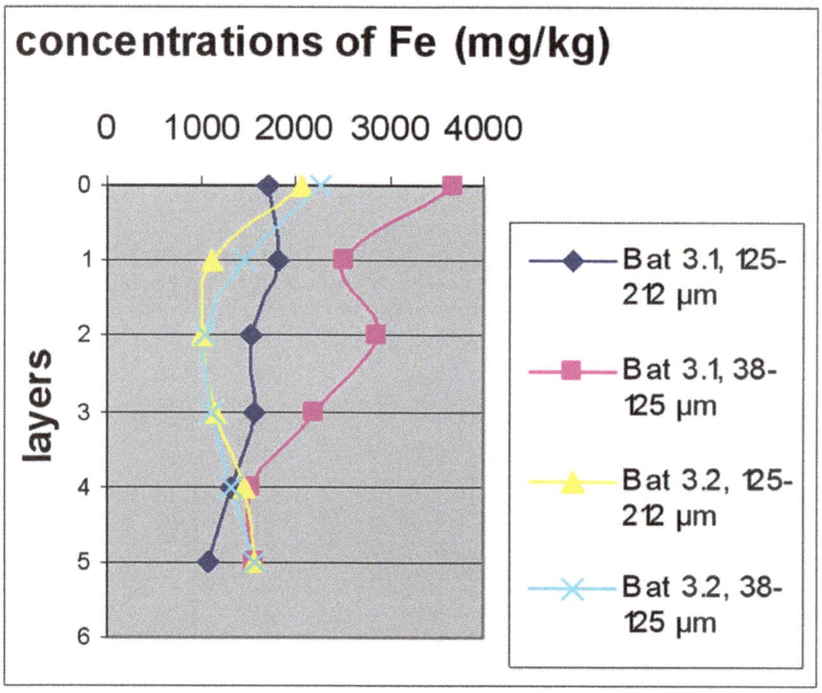

(

f)

Figure 7. Change in the concentration of Fe (mg/kg) according to the layers for both size fractions in the two sites Beirut and Batroun in B 0m (a), B 5m (b), B 10m (c), Bat 1 (d), Bat 2 (e), et Bat 3 (f).

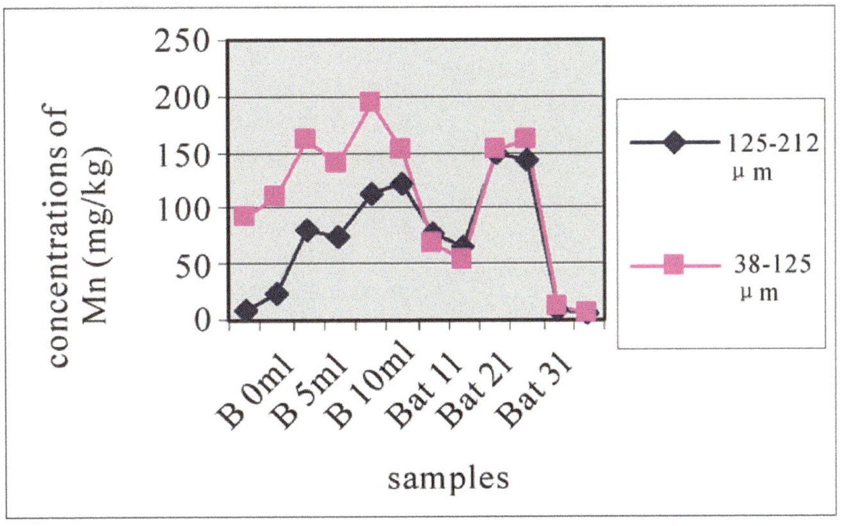

(a)

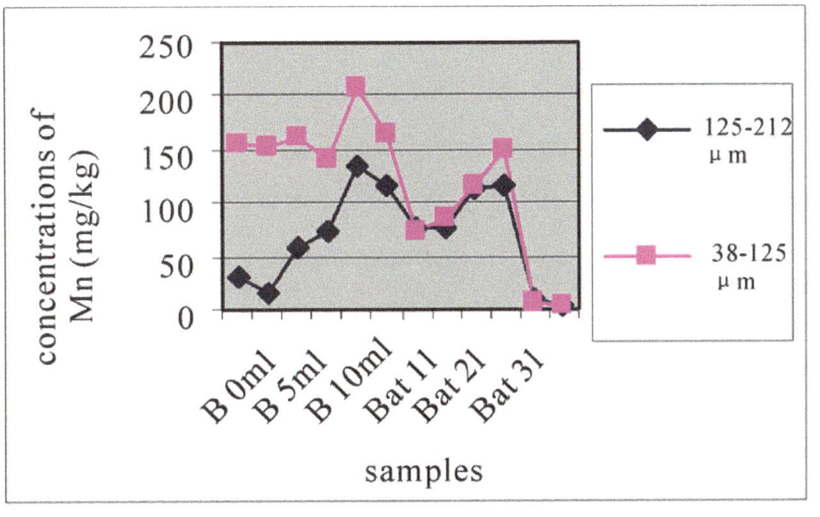

(b)

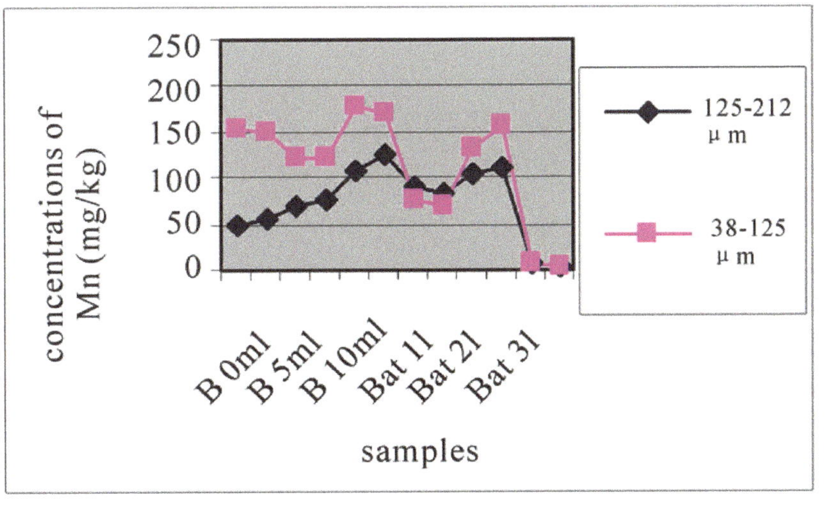

(c)

Figure 8. Change in the concentration of Mn for the both size fractions of the first layer (a), the seventh layer (b) and the eleventh layer (c), in the different levies of the two sites.

For the size fraction 38-125 μm, **Figure 10** shows that the site of Batroun was most concentrated; this was the current state of contamination of the study area. In **Figure 10**-b and 10-c, both sites were polluted, but pollution in Beirut is merely higher than in Batroun. These two figures represent the state of contamination and temporal evolution of the sites. Bat 3, B 0m and B 10m were most polluted in Cu than the other station.

Figure 11 shows the change in the concentration of Cu according to the different layers of a levy. The different figures show that there was no a large difference according to different layers. It is clear from **Figure 10** and 11 that there was no significant difference for the repetition of the majorities of the samples. Samples with the size fraction finer (38-125 μm) had higher concentrations compared to the other portion size of 125-212 μm.

Concentration of Pb

Compared with the concentration of Pb in samples taken from Jiyeh (17 μg/l) [5,6], from Zouk and Selaata (2μg/l), it was resulted that the concentration of lead has increased.

Figures 12(a), (b) and (c) reveal the change in the concentration of Mn, each one concerned a specific layer, between the different sampling made in the two sites and for the two size fractions selected before. Different fig ures show that Batroun site was more concentrated in Pb than Beirut site. This difference was more appreciable in the first layer, which was represented the state of a current and recent contamination. Bat 3 and B 10 and B0m were more concentrated in Pb than the other one.

Figure 13 shows the change in the concentration of Cu according to the different layers of a levy. Figures 12 and 13 show that for the site of Beirut, there was no significant difference between layers for the same depth, while for the site of Batroun, there was some difference between layers of the same depth.

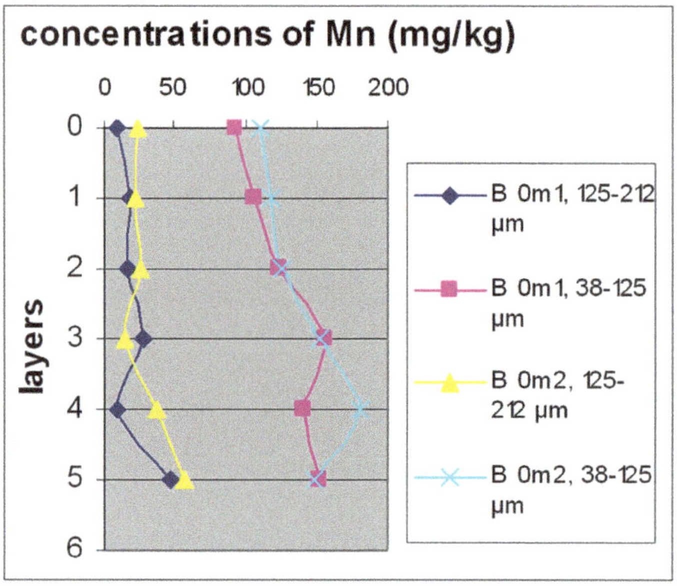

(a)

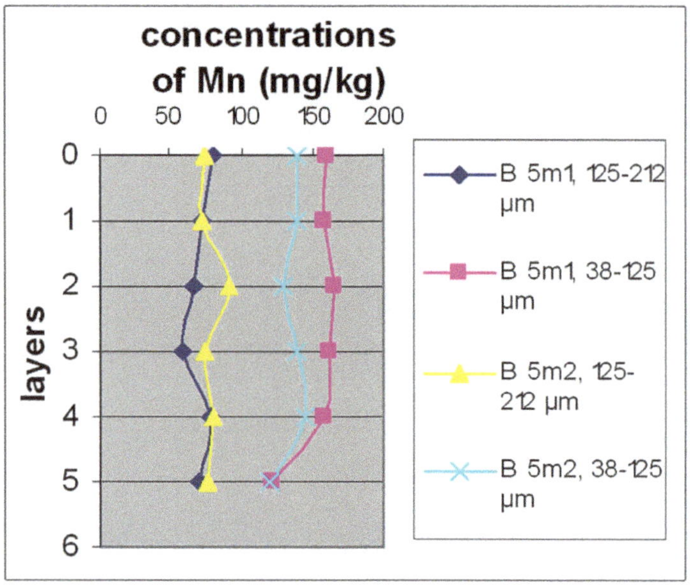

(b)

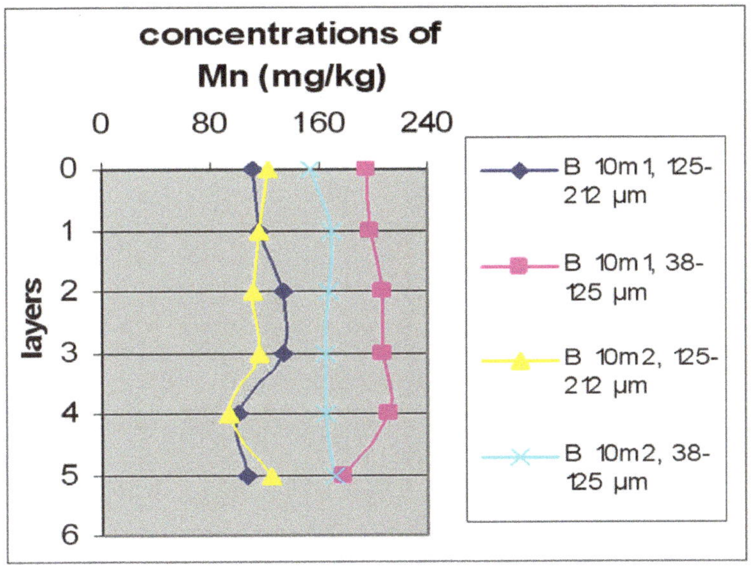
concentrations of Mn (mg/kg)
0
80
160
240
layers
0
1
2
3
4
5
6
B 10m1, 125-212 µm
B 10m1, 38-125 µm
B 10m2, 125-212 µm
B 10m2, 38-125 µm

(c)

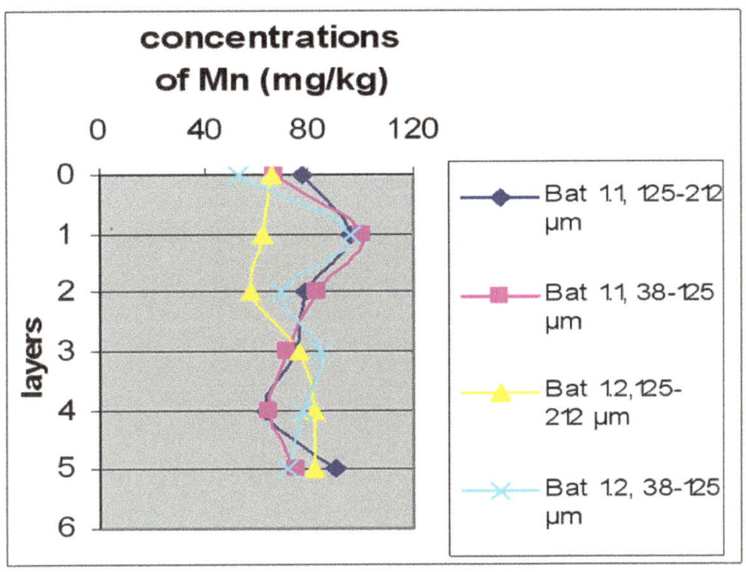
concentrations of Mn (mg/kg)
0
40
80
120
layers
0
1
2
3
4
5
6
Bat 1.1, 125-212 µm
Bat 1.1, 38-125 µm
Bat 1.2, 125-212 µm
Bat 1.2, 38-125 µm

(d)

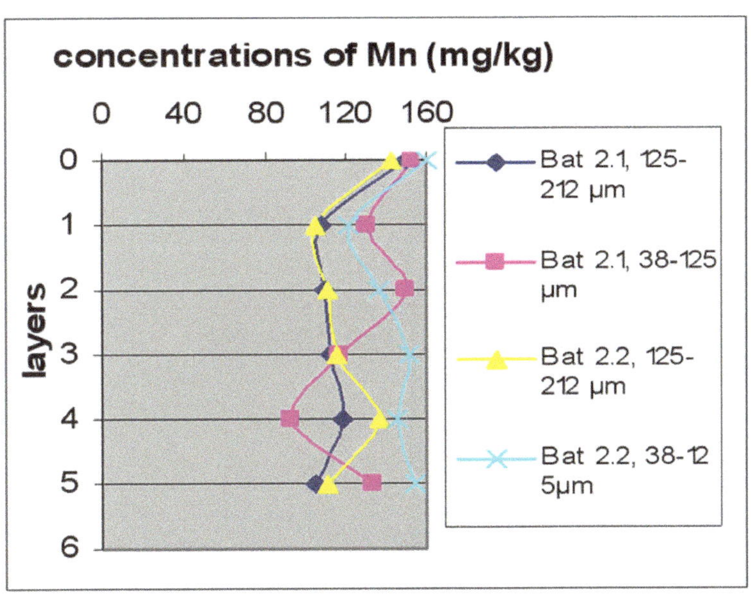
concentrations of Mn (mg/kg)
0
40
80
120
160
layers
0
1
2
3
4
5
6
Bat 2.1, 125-212 µm
Bat 2.1, 38-125 µm
Bat 2.2, 125-212 µm
Bat 2.2, 38-125µm

(e)

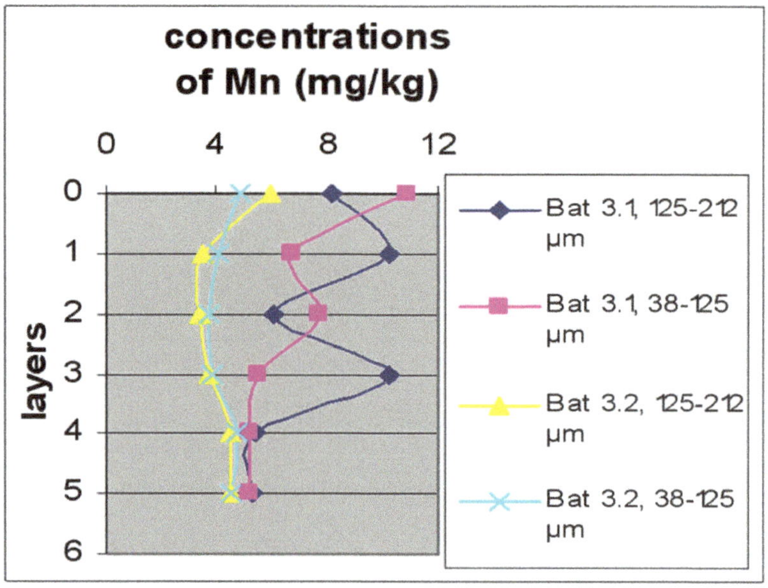

f)

Figure 9. Change in the concentration of Fe (mg/kg) according to the layers for both size fractions in the two sites Beirut and Batroun in B 0m (a), B 5m (b), B 10m (c), Bat 1 (d), Bat 2 (e), et Bat 3 (f).

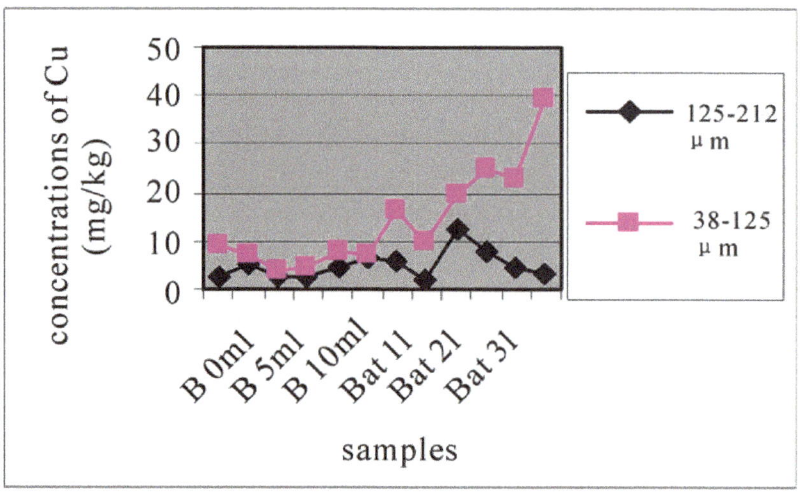

(a)

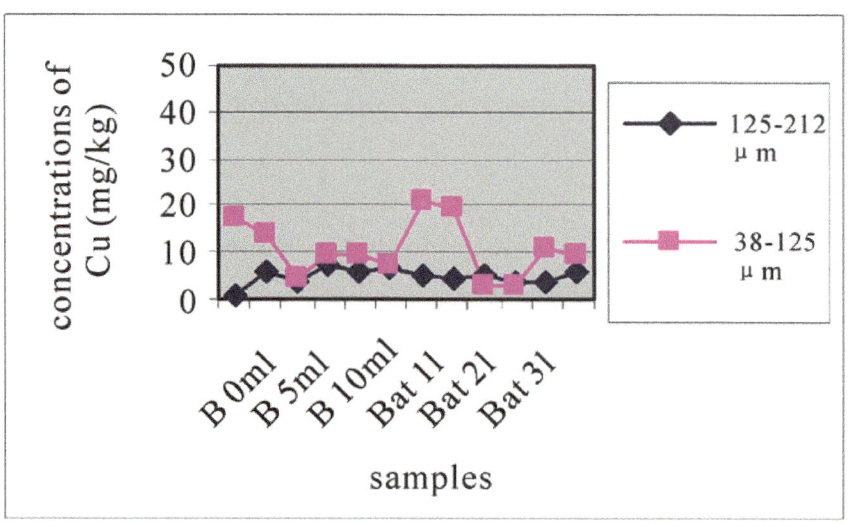

(b)

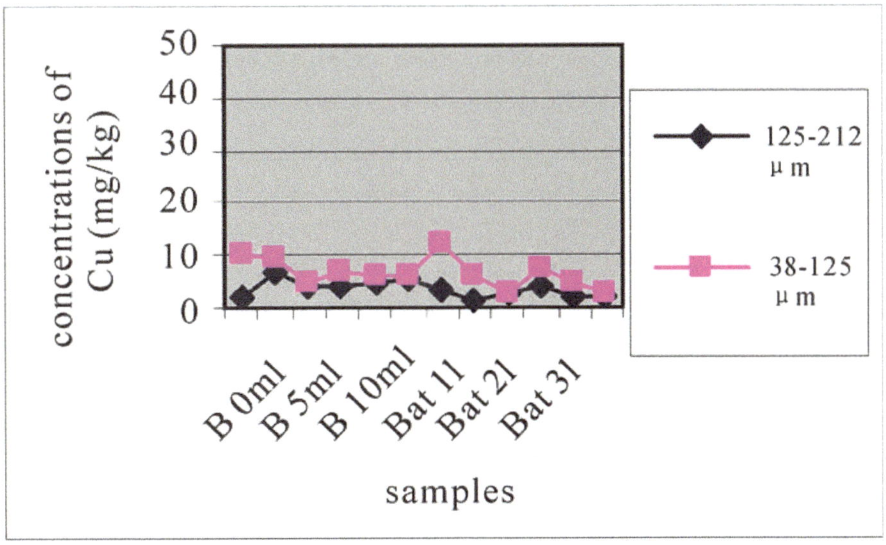

(c)

Figure 10. Change in the concentration of Cu for the both size fractions of the first layer (a), the seventh layer (b) and the eleventh layer (c), in the different levies of the two sites.

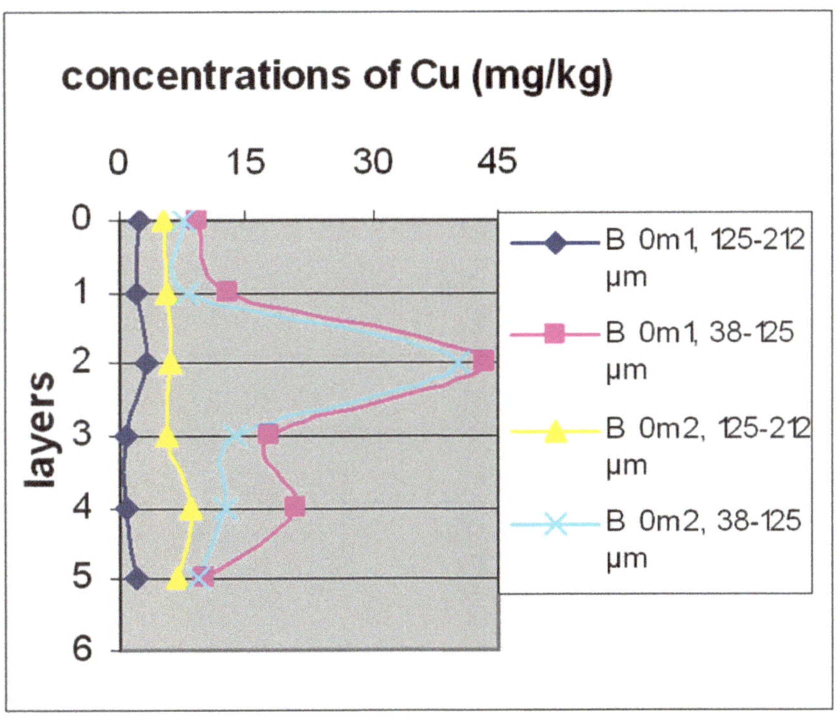

(a)

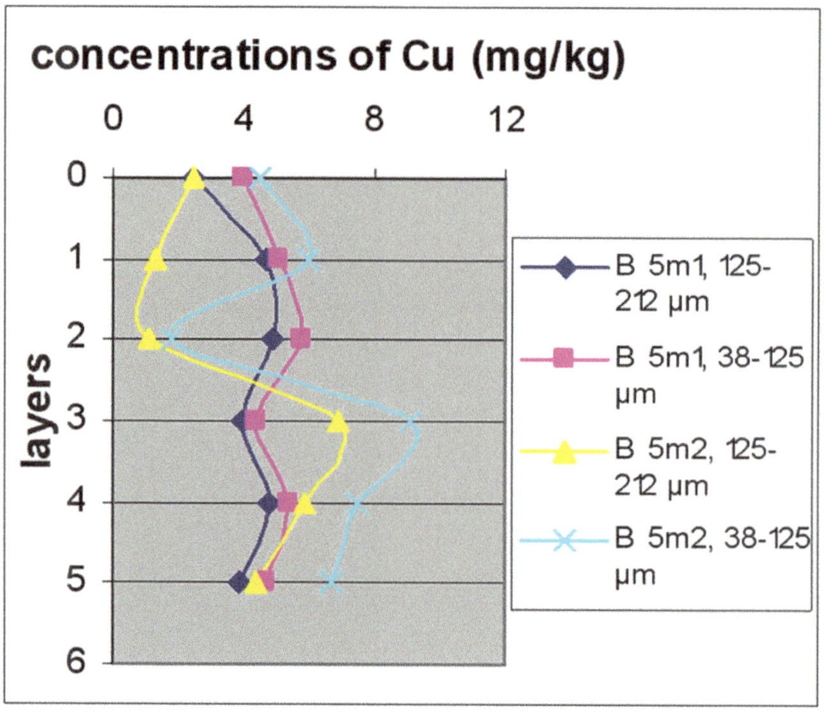
concentrations of Cu (mg/kg)
0
4
8
12
layers
0
1
2
3
4
5
6
B 5m1, 125-212 µm
B 5m1, 38-125 µm
B 5m2, 125-212 µm
B 5m2, 38-125 µm

(b)

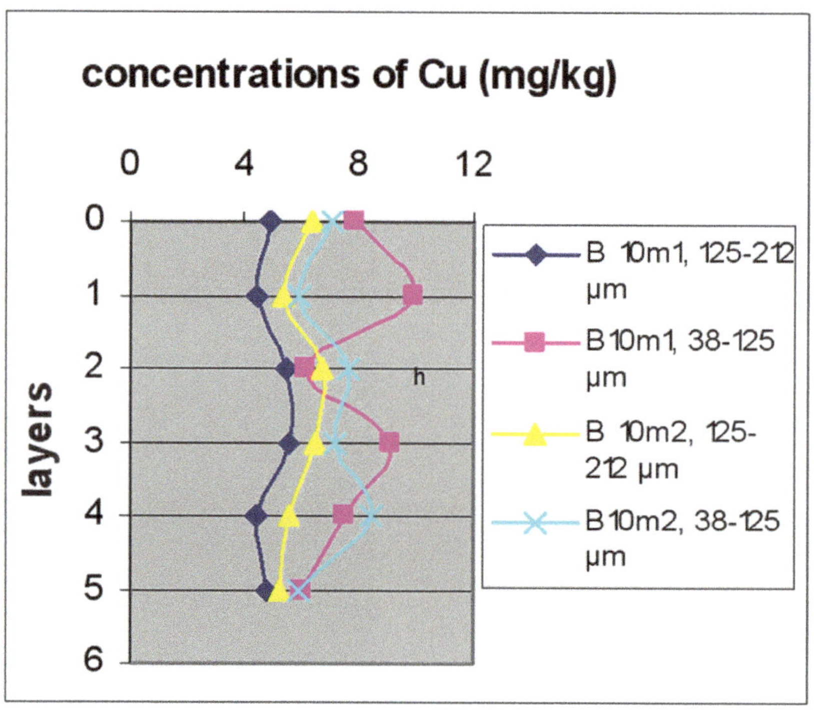
concentrations of Cu (mg/kg)
0
4
8
12
layers
0
1
2
3
4
5
6
B 10m1, 125-212 µm
B10m1, 38-125 µm
B 10m2, 125-212 µm
B10m2, 38-125 µm

(c)

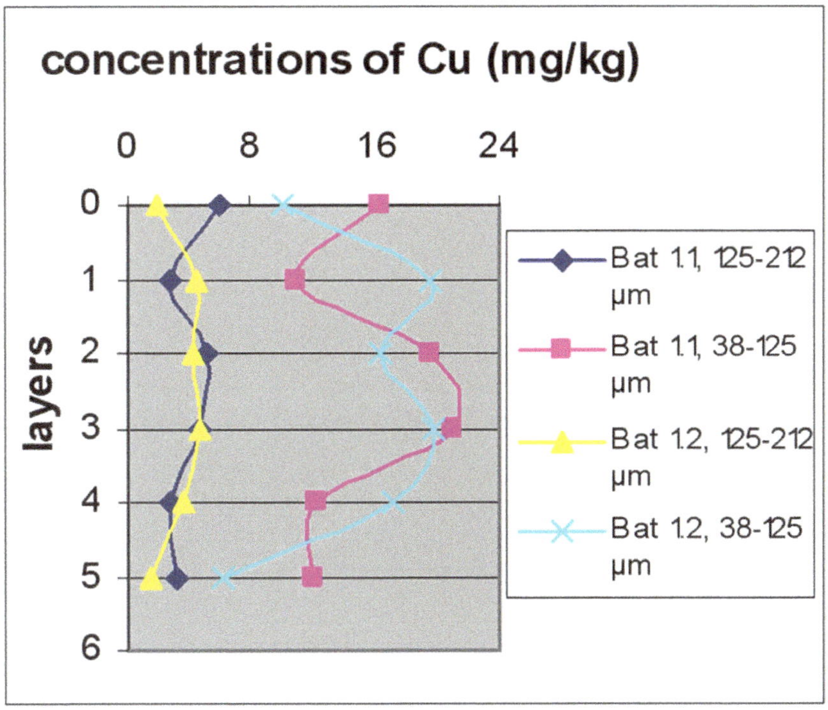
concentrations of Cu (mg/kg)
0
8
16
24
layers
0
1
2
3
4
5
6
Bat 1.1, 125-212 µm
Bat 1.1, 38-125 µm
Bat 1.2, 125-212 µm
Bat 1.2, 38-125 µm

(d)

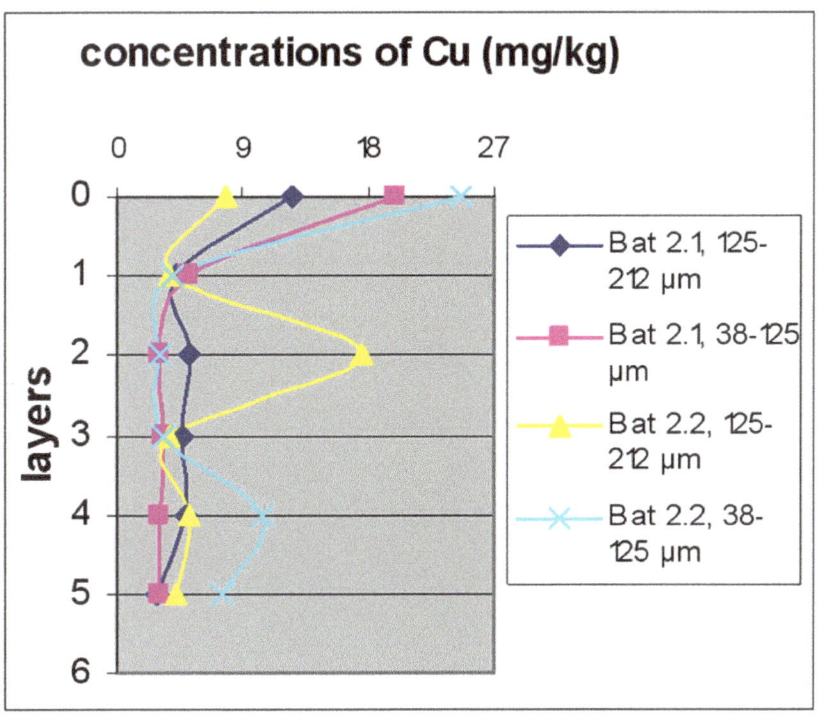
concentrations of Cu (mg/kg)
0
9
18
27
layers
0
1
2
3
4
5
6
Bat 2.1, 125-212 µm
Bat 2.1, 38-125 µm
Bat 2.2, 125-212 µm
Bat 2.2, 38-125 µm

(e)

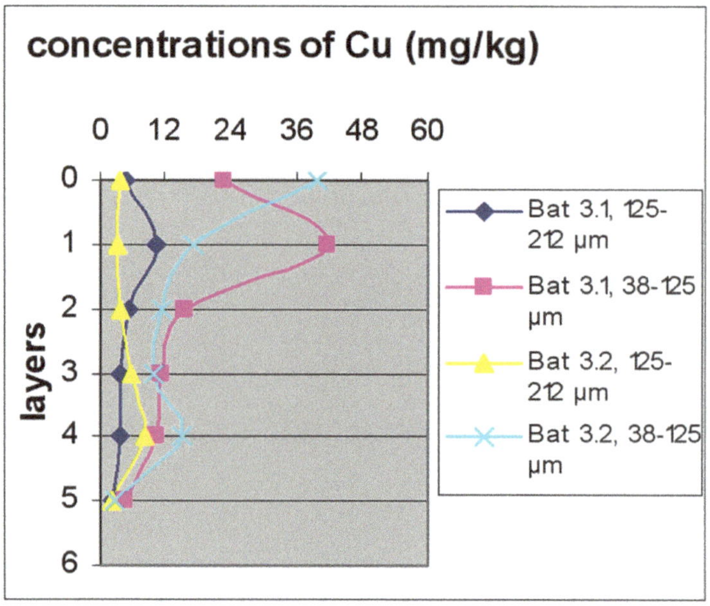

(f)

Figure 11. Change in the concentration of Fe (mg/kg) according to the layers for both size fractions in the two sites Beirut and Batroun in B 0m (a), B 5m (b), B 10m (c), Bat 1 (d), Bat 2 (e), et Bat 3 (f).

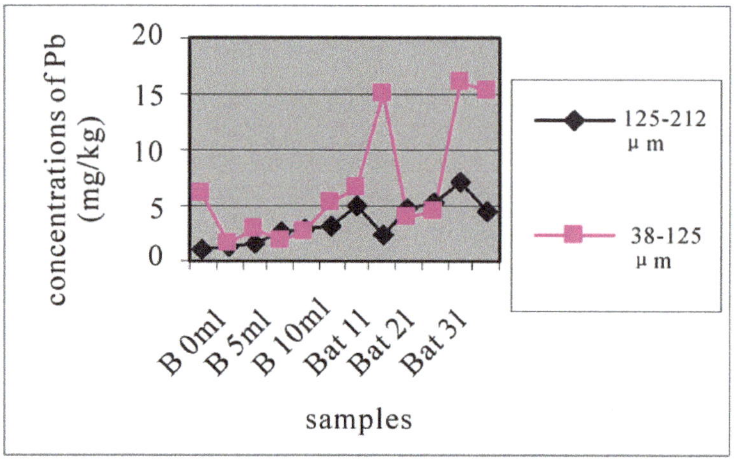

(a)

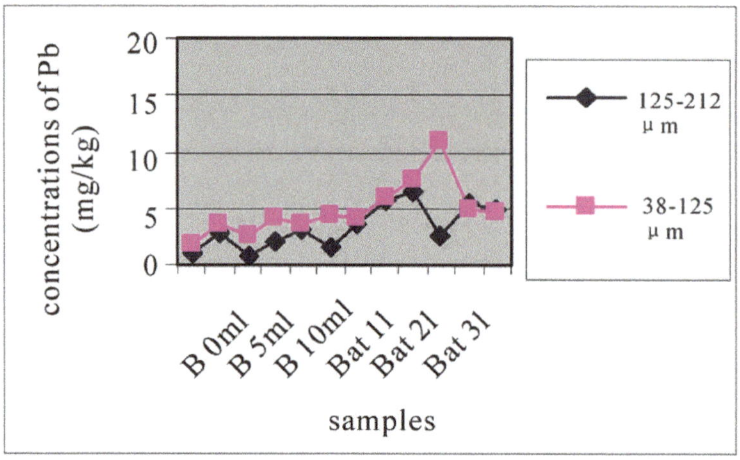

(b)

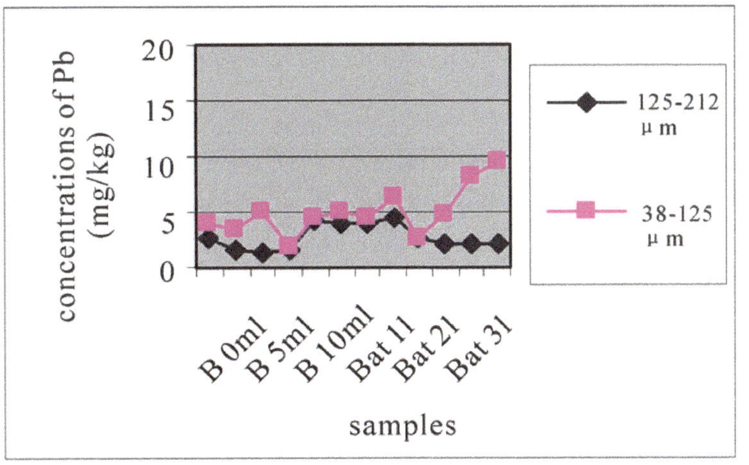

(c)

Figure 12. Change in the concentration of Cu for the both size fractions of the first layer (a), the seventh layer (b) and the eleventh layer (c), in the different levies of the two sites.

The difference between the layers in the site of Batroun can be attributed to the lack of disruption during the accumulation of pollutants over time, allowing mixing contaminants.

Almost all charts in **Figure 13** show that repetition did not make a significant difference. The analyses also show that samples with finer size fraction show slightly higher concentration of Pb over the other portion size.

Analysis of Principal Component (A C P)

All parameters in this study show that the axis F1 (which represent 45% of the variance) was mainly influenced by Fe and Mn, and the axis F2 (which represents 37% of the variance) was primarily influenced by Cu and Pb (**Figure 14**). This analysis also shows that Fe and Mn are perfectly correlated with each other positively and negatively with Cu and Pb, allows dividing the two heavy metal Cu and Pb one the one hand and the two other metal Fe and Mn on the other hand to evaluate the sites.

This is widely visible in the table of Person correlation (**Table 5**). This correlation indicates the presence of a linear relationship between Fe and Mn, and another linear relationship between Cu and Pb, because Fe and Mn were considered as "Siderophile" element, but Cu and Pb were considered as "Chalcophile" element. For a longterm monitoring, the study can focus on a single parameter instead of two.

A C P made on both parameters (Pb and Cu) in Figures 14 and 15 shows that the correlation between Cu and Pb is significant and there was a significant difference between the two selected sites.

Conclusions

The comparison of the concentrations of Fe, Mn, Cu and Pb, obtained in this study, according to standards [3,15, 16] indicated that they did not exceed the maximum limit with the exception of a few values of Fe.

However, the values are far greater than those existed before the conflict of 2006, due to the release of fuel oil into the sea. The comparison with the studies done before the conflict is limited since no studies have addressed different layers of sediment. This is because the values given in these studies return to sediment sample with diameters of less than 63 or 80 µm [6,18].

It was noticeable that there is no correlation between Fe and Mn and between the Cu and Pb, thus the parameters are divided in two groups: a-Cu and Pb, bFe and Mn.

For Fe and Mn, sediments of Beirut site (B 10 m station is the most concentrated), are more concentrated than in Batroun site with the exception of Bat 2, which contained a high concentration of Fe and Mn. For Cu and Pb, Batroun site had higher concentration than Beirut site in the surface layers. These results are opposed to those found before the conflict of 2006. The high concentration in Beirut site is probably due to human activities and the conflict of 2006 (oil spill in the sea).

The high concentration in Batroun site was attributed to the impact of wind carried pollutants from different

Table 5. Correlation of pearson between the heavy metals studied

	Cu	Pb	Fe	Mn
Cu	1	**0.474**	0.091	0.028
Pb	**0.474**	1	-0.099	-0.145
Fe	0.091	-0.099	1	**0.753**
Mn	0.028	-0.145	**0.753**	1

The significant values are represented in bold with α=0. 05

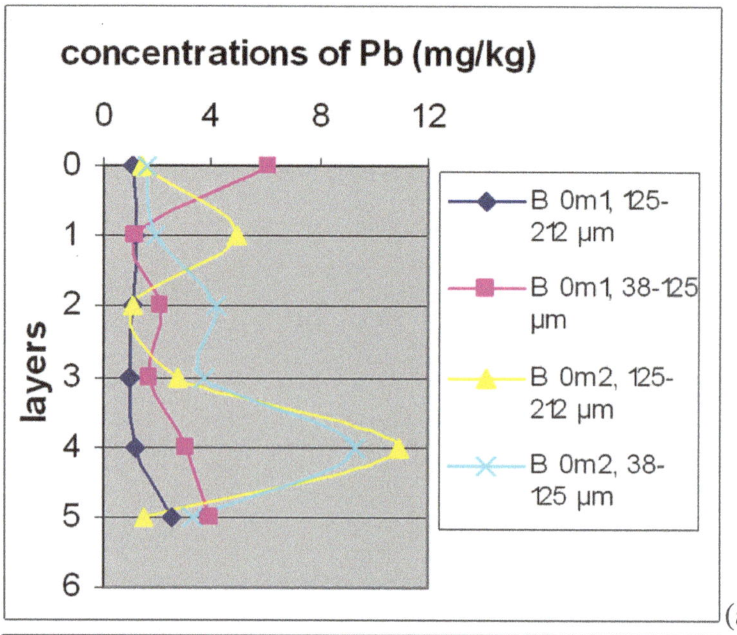
concentrations of Pb (mg/kg)
0
4
8
12
0
1
2
3
4
5
6
layers
B 0m1, 125-212 µm
B 0m1, 38-125 µm
B 0m2, 125-212 µm
B 0m2, 38-125 µm

(a)

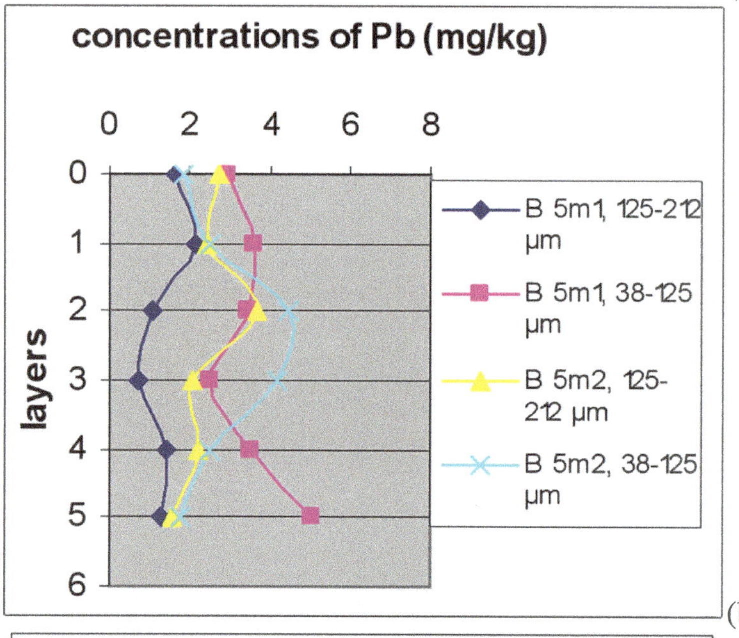
concentrations of Pb (mg/kg)
0
2
4
6
8
0
1
2
3
4
5
6
layers
B 5m1, 125-212 µm
B 5m1, 38-125 µm
B 5m2, 125-212 µm
B 5m2, 38-125 µm

(b)

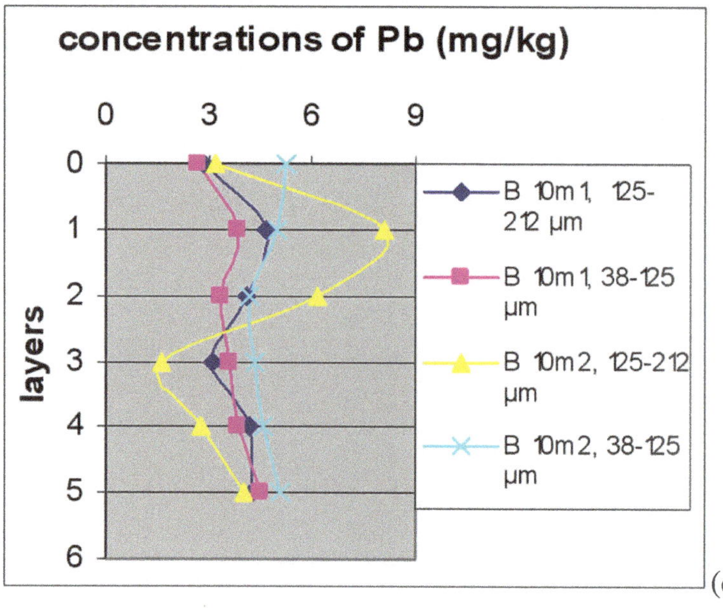
concentrations of Pb (mg/kg)
0
3
6
9
0
1
2
3
4
5
6
layers
B 10m1, 125-212 µm
B 10m1, 38-125 µm
B 10m2, 125-212 µm
B 10m2, 38-125 µm

(c)

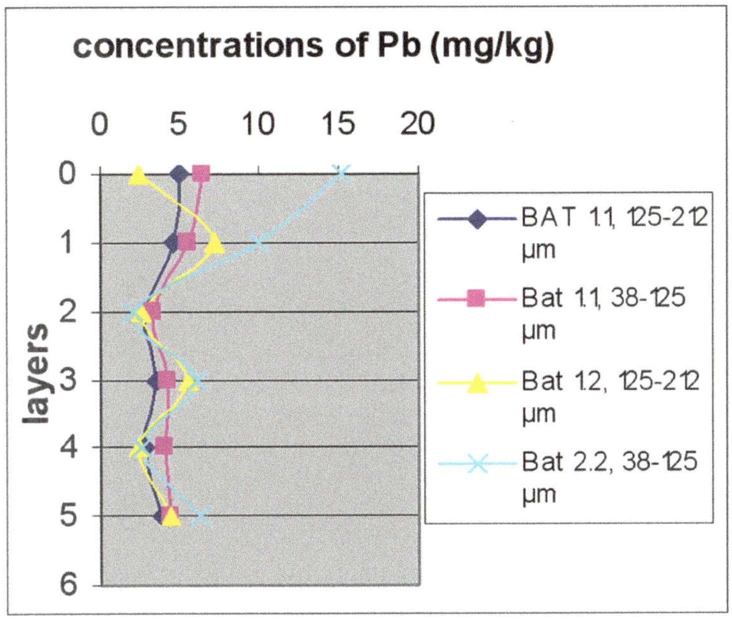

(d)

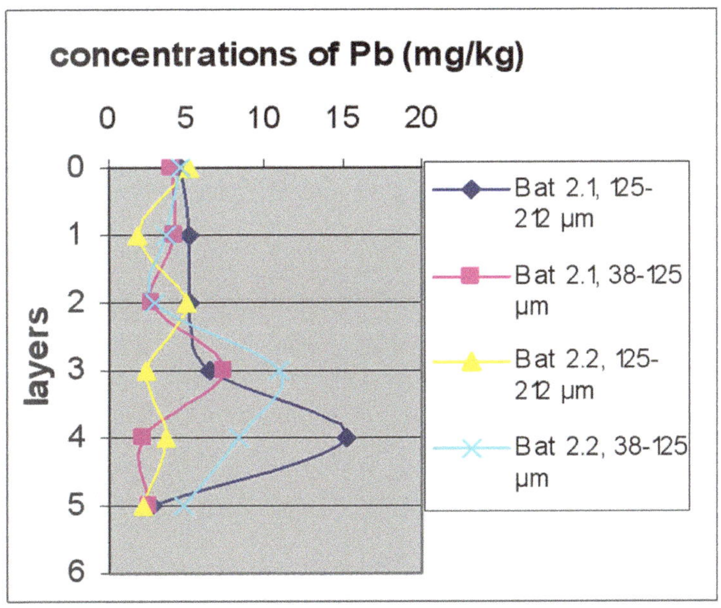

(e)

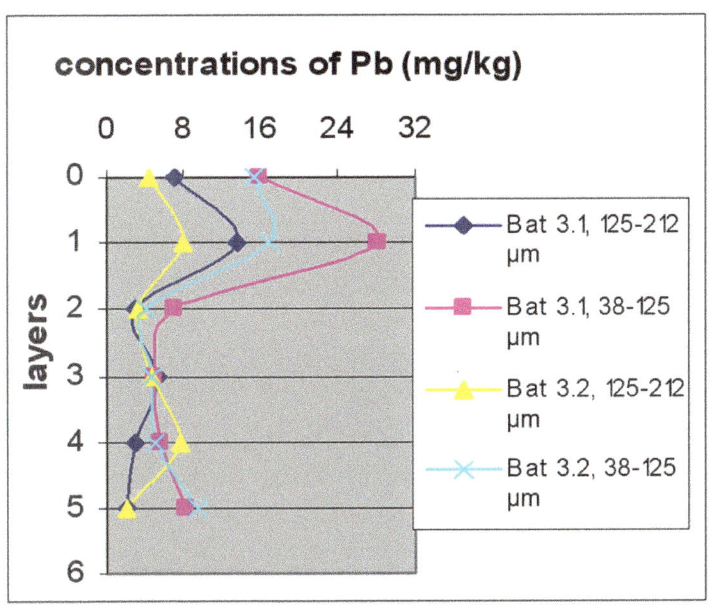

(f)

Figure 13. Change in the concentration of Fe (mg/kg) according to the layers for both size fractions in the two sites Beirut and Batroun in B 0m (a), B 5m (b), B 10m (c), Bat 1 (d), Bat 2 (e), et Bat 3 (f).

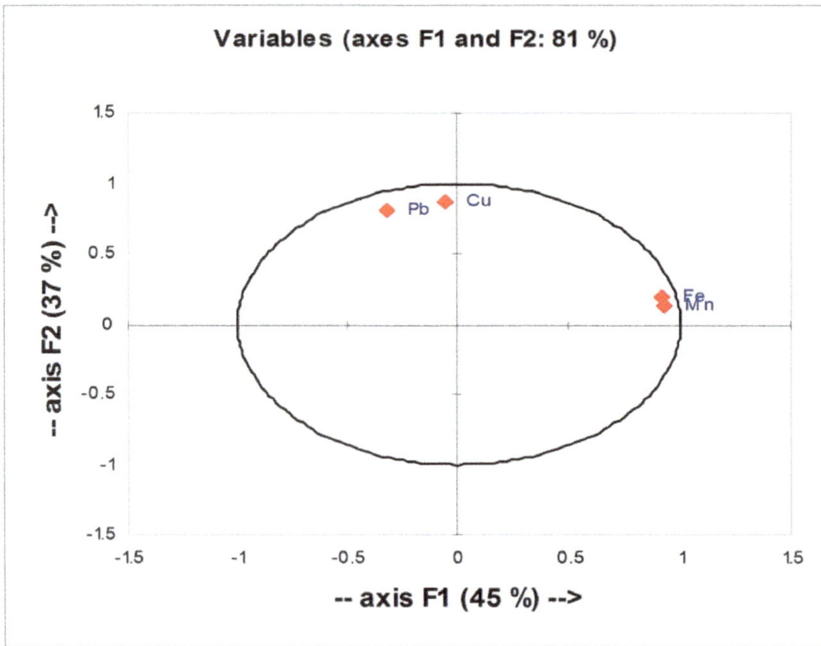

Figure 14. Analysis of principal component (ACP) to all parameters studied (in Beirut and Batroun): correlations between concentrations of different heavy elements traces in all samples studied.

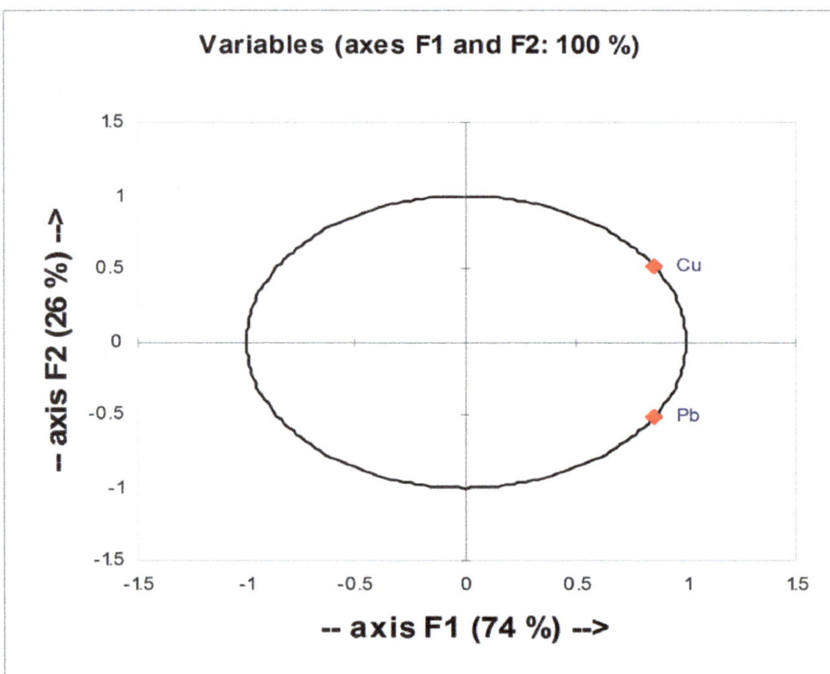

Figure 15. Analysis of principal component (ACP) to all parameters studied (in Beirut and Batroun) : correlations between concentrations of Cu and Pb in all samples studied.

regions, and the correlation between heavy metals and hydrocarbons, was attributed to the industry of chemical products at Selaata factory near Batroun, as well as due to the conflict of 2006.

For the investigated parameters, the effect of size fraction is significant, because the samples with the finer size fraction 38-125 µm had higher concentration than the other fraction 125 – 212 µm. The mineralogical composition was also involved in significant proportions.

Thus, the silty clay minerals were more richer in metals that silica sand. This is the case with stations B 10m, Bat 2 and Bat 3 which contained quantities exceeding of fine fractions, because results showed higher concentrations of metals (especially Fe and Mn) compared to other stations [18–20].

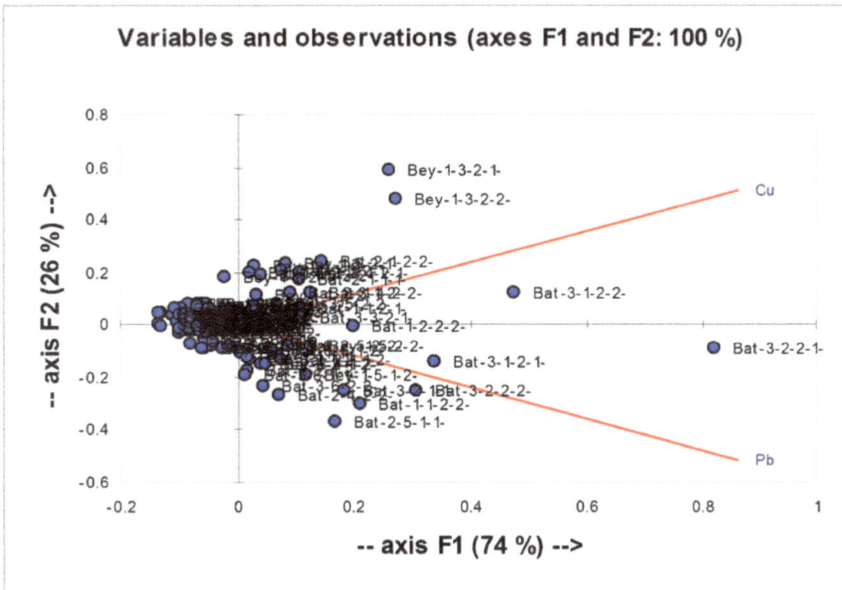

Figure 16. Analysis of principal component (ACP) concerning the concentration of Cu and Pb to all parameters studied (in Beirut and Batroun).

For all studied parameters the effect of depth was significant, and it shows that pollution is growing more; especially with depth at a distance from the coast, but also at the coast which was close to industries, traffic and various sources of pollution.

Similarly, the effect of repetition (2 levies distant from one another by 30 cm) was not significant for different studied metals; proving that for the same sample taken on the same depth, there was no difference, which also shows that the variance linked to the analysis, is void.

For the investigated parameters, there was no effect between layers except for Pb. This is in accordance with the results of the variance analysis, while in the future we can study a single layer for daily analysis in the same region.

It sought to continue this research on much more sites and on time series in order to confirm the effect of layers in other regions along the Lebanese coast.

REFERENCES

1. R. Steiner, "Lebanon oil spill rapid assessment and response mission," pp. 1–30, 2006.
2. E. Iaurif, "Regional environmental assessment, report on the coastal zone of Lebanon," Government of Lebanon, Council for Development and Reconstruction, pp. 46–56, 1997.
3. R. Kantin and G. et Pergent, "Gestion des écosystemes littoraux méditérranées," IFREMER et Université de CORSE, pp. 15–45, 2007.
4. S. Casas, "Modélisation de la bioaccumulation de métaux traces (Hg, Cd, Pb, Cu et Zn) chez la moule Myitlis Galloprovinciallis, en milieu méditerranéen," Sud Toulon VAR. France, pp. 10–19, 2005.
5. K. Nakhle and Le Mercure, "le Cadmium et le Plomb dans les eaux littorales libanaises: apports et suivi au moyen de bioindicateurs quantitatifs (éponges, bivalves, et gastropodes)," Paris 7. France, pp. 7–14, 32–41, 99–101, 2003.
6. N. Nassif, "Pollutions chimiques en milieu marin : Essai de modélisation et approche réglementaire," Edition: GG/F. Janvier, Institut National Agronomique. Paris-
7. Grignon. Paris-France. ISBN: 2–35040–000–X, pp. 9–42, 81, 203–210, annexes 13, 14, 2006.
8. J. C. Amiard, "Les problèmes liés à l› échantillonnnage et à la détection des élements traces en écotoxicologie," Vol. 1, pp. 172–195, 1994.
9. M. Castrec-Rouelle, N. Nassif, and A. M. et DeKersabiec, "Prélèvements préparation et traitement d›échantillons environnementaux: Approches statistiques," Dossier Environnement, Vol. 32, pp. 23–28, 2003.
10. C. Biney, A. T. Amuzu, D. Calamari, N. Kaba, I. L. Mbome, H. Naeve, O. Ochumba, O. Osibanjo, V. Radegonde, and M. A. H. et Saad, "Etude des métaux lourds," Revue de la pollution dans l›environnement aquatique africain, FAO, Vol. 25, pp. 1–6, 1994.
11. B. Boutier, D. Claisse, D. Auger, E. Rozuel, J. Breteau-
12. deau, and I. Truquet, "Surveillance du milieu marin. Les métaux dans les sédiments du Golfe de Gascogne," RNO (IFREMER), pp. 17–34, 2005.
13. E. Bastarache, "Toxicologie, céramique, verrerie et métallurgie," Smart. Conseil, 2006.
14. PNUE, "Meilleurs méthodes de gestion pour les sources agricoles non ponctuelles de pollution," Rapport techni-
15. que du PEC, Programme pour l›environnement des caraibes, pp. 150, 1998.
16. R. Mebazaa, "Détermination de la taille de l›échantillon des sédiments marins par étude de la variance," L›Institut National Agronomique Paris-Grignon - Centre National des Sciences Marines de Jounieh. Agence Universitaire de la Francophonie (AUF). Mémoire de Diplôme d›études approfondies, Contrôle et gestion de la qualité, pp. 12–22, 2004.
17. PNUE, "Etat du milieu marin et littoral de la région méditérranéene," MAP Technical Reports-Series, pp.

3–11, 29–48, 1996.

18. UNEP, "Etat de l›environnement et politiques suivies de 1972 A 2002, la mer et les côtes. GEO 3," Le passé, le présent et les perspectives d›avenir, pp. 180–209, 2002.
19. B. Boutier and D. et Claisse, "Surveillance du milieu marin. Les contaminants chimiques dans les sédiments du littoral méditérranéen," RNO (IFREMER), pp. 9–52, 1998.
20. B. Boutier and D. et Claisse, Surveillance du milieu marin. Les carottes sédimentaires, mémoire de la contamination," RNO (IFREMER) , pp. 21–40, 2001.
21. J. F. Chiffoleau, D. Auger, B. Boutier, E. Rozuel, and I. et Truquet, "Dosage de certains métaux dans les sédiments et les matières en suspension par absorption atomique," IFREMER, pp. 6–16. 2003.
22. Y. Gueguen, "Réalisation d›un système expert pour le bilan de la contamination métallique du réseau hydrographique de la Seine," Université Pierre et Marie Curie, Université Paris-Sud, et Ecole Nationale du Génie Rural des Eaux et des Forêts. France, pp. 5–6, 2003.
23. B. Boutier, D. et Claisse, "Surveillance du milieu marin. Les métaux lourds dans les sédiments de la Baie de Seine (campagne 1993)," RNO (IFREMER), pp. 25–32, 1995.

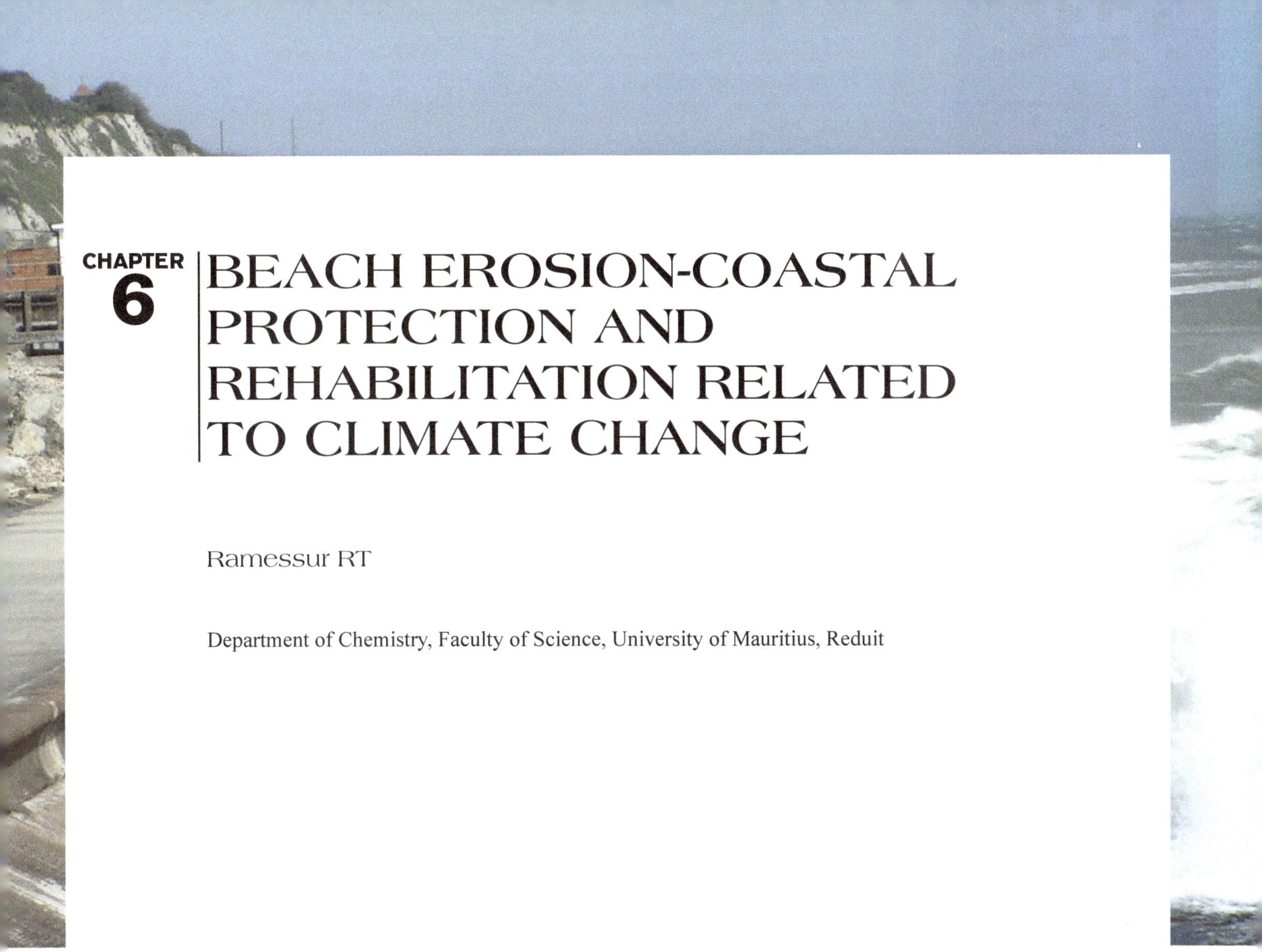

CHAPTER 6

BEACH EROSION-COASTAL PROTECTION AND REHABILITATION RELATED TO CLIMATE CHANGE

Ramessur RT

Department of Chemistry, Faculty of Science, University of Mauritius, Reduit

INTRODUCTION

Long term data (1967 to 2012) available by aerial photographs and satellite images showed that about 11% of beaches around Mauritius were eroded, 30% accreted and 59% were stable. Percentage of living corals in the **lagoon** has been estimated to be 27% with a decrease from 50% (2001) to 27% (2012). Three coasts namely Albion, Bel Ombre and Mon Choisy have been severely eroded with most significant beach erosion taking place at Albion (-0.5m/year). The application of indicators in Integrated Coastal Zone Management (ICZM) and **sustainable development** in SIDS including Mauritius has become essential [1-4]. While environmental indicators have long been used to monitor the state of the coastal and **marine environment**, socioeconomic indicators have seen very limited application, and governance indicators have been applied mainly in reporting of the management process. A great challenge lies in using appropriate sets of governance and socioeconomic indicators that will allow decisionmakers to determine whether ICZM interventions are achieving their intended goals such as beach rehabilitation against erosion [5-12]. The application of indicators in **Integrated Coastal Zone Management** (ICZM) to monitor the state of the coastal and marine environment together with **socioeconomic** indicators and governance indicators in relation to climate change is required. Development of ICZM related to climate change and coastal engineering online modules suit well in the constructivist context as it is of sharp contrast with behavioral approach where the instruction pathway is guided. The ICZM and coastal engineering learner-centered technique and the interactions between people in ICZM networked/mobile learning environments in synchronous/asynchronous mode can be useful. The interactions through text, audio, graphics, video, **coastal modeling**, shared workspaces or a combination of these forms are useful as networked learning is interactive, flexible and promotes active engagement as well as aiding reflective learning practice [13-15]. These attributes of networked learning for coastal engineering, ICZM and climate change learner-centred instruction framework is also being developed. Other objective can be to assess and compare aspects of biogeochemical cycles of some dissolved nutrients and trace metals in coastal waters, submarine groundwater discharge, estuarine sediment contamination and related to beach erosion. Studies now currently focus on dissolved nutrients (nitrate and phosphate) and some selected trace metals Cu, Zn and Pb along the **coastline**. The concentration of trace metals in both reduced and oxidised surface sediments and seasonal events during summer and winter which may lead to significant changes in trace metal deposition and dissolved nutrients is being studied and ligands which exhibit high selectivity for metal ions such

Citation: Ramessur RT (2016) Beach Erosion-Coastal Protection and Rehabilitation related to Climate Change . J Coast Zone Manag 19: e117. doi:10.4172/jczm.1000e117

as copper (II) for wastewater treatment is being investigated. Twelve beaches around Mauritius have also been preselected for development of coastal conservation plans. Two sites were selected for implementation of demonstration projects and 5 sites for monitoring purposes. The two demonstration project site comprises one where Physical Intervention (PI) can be carried out and one where there would be No Physical Intervention (NPI). The criteria for selection of a specific site included natural and environmental conditions, obstacles to land use, importance for protection of hinter area, validity for future development plan, and national policy among others under the JICA Project "Capacity Building in Coastal Protection and Rehabilitation in the Republic of Mauritius" [16-19].

REFERENCES

1. Ramessur RT (2013) A Review of Coastal Zone Management Facing Climate Change and Natural Disasters in Mauritius. J Geogr Nat Disast 1: 003
2. Ramessur RT, Gunputh RP, Ramessur TS (2013) Climate Compatible Development: Legal Implications in the Coastal Zone and Inclusive Development for Mauritius. J Coast Dev 16: 378.
3. Ramessur RT (2013) Coastal Zone Facing Climate Change: Knowledge Towards New Solutions. J Coast Dev 16: 103.
4. Ramessur RT (2013) Coastal Morphodynamics and Human Impacts. J Coast Dev 16: 104.
5. Ramessur RT and Boodhoo K (2013) Trace Metal Contamination in Urban Sediments in Port Louis, Mauritius. Ecology and Environment Research 1: 155-160
6. Ramessur RT, Boodhoo K, Balgobin J, Povinec P, Burnett WC (2012) Dissolved nutrients in submarine groundwater discharge in Flic en Flac lagoon, Mauritius. Sixth Wiomsa Symposium. Western Indian Ocean Journal of Marine Science 10: 121-127
7. Povinec PP, Burnett WC, Beck A, Bokuniewicz H, Charette M, et al. (2012) Isotopic, geophysical and biogeochemical investigation of submarine groundwater discharge: IAEA-UNESCO intercomparison exercise at Mauritius Island. Journal of Environmental Radioactivity 104: 24-45.
8. Boodhoo K, Ramessur RT (2015) Exploration and Potential Exploitation of Deep Sea Minerals in the EEZ and ECS of Mauritius. Proceedings International Workshop on Minerals Processing and Beneficiation, Johannesburg, South Africa: 227-240.
9. Roshan TR (2015) Integrated Coastal Zone Management: Ecological, Socio- Economic and Governance Challenges. J Coast Zone Manag 18: 112.
10. Lollchund MR, Dilmahamod F, Boodhoo K, Ramessur RT (2015) Correlations between monthly precipitation, SST, humidity and CO2 levels in the tropical South West Indian Ocean (SWIO) basin. RESUS 15 Proceedings.
11. Ramessur RT (2009) Networked Learning for an Online ICZM module. Proceedings 20 th ICCE – In: Chemistry in the ICT Age: 460.
12. Swayne D, Denko M, Fagoonee I, Marshall L, Molina G (2007) ICT Issues for Environmental monitoring and management. Chapter 6: Environment. IFIP WITFOR 2007 (World Information Technology Forum). IFIP Press Austria: 97-110.
13. Ramessur RT Santally MI (2007) Computer mediated communication for effective teaching-learning of coastal zone management module. International Journal of Education and Development 3: 1.
14. Ramessur RT, Parry SJ (1998)Variation of major elements and nutrients in a coastal system in Mauritius. University of Mauritius Research Journal 2: 145-157.
15. Ramessur RT (1997)Assessment of a tropical lagoon in Mauritius using semi-quantitative techniques. Rev Agri et Suc 76: 26-30.
16. Ramessur RT (1996)The use of a GIS for the coastal zone management of a seaside resort area in Mauritius. PROSI 324: 20-25.
17. Ramessur RT (1991)Coral zonation at Flic en Flac. University of Mauritius Research Journal: 3-11 .
18. Ramessur RT (2014) Sustainable Development: The Need for Vulnerability Resilience Profile and the Green Economy Transition in Mauritius. J Geogr Nat Disast 4:116.
19. Ramessur RT (2015) Ocean Economy-Regional Climate Modelling and Sustainability Issues. J Coast Zone Manag 18:115.

CHAPTER 7

YELLOWFIN SEABREAM, A TARGET SPECIES FOR MONITORING OF COASTAL POLLUTANTS

Aliakbar Hedayati

Faculty of Fisheries and Environment, Gorgan University of Agricultural Sciences and Natural Resources, Gorgan, Iran

FISH SPECIES FOR ECO-TOXICOLOGICAL STUDIES

For several reasons, fish species have attracted considerable interest in studies assessing biological and biochemical responses to contaminants. Despite their limitations, such as a relatively high mobility, fish are generally considered to be the most feasible organisms for pollution monitoring in aquatic systems. Fishes are by far the most diverse class of vertebrates; the 28,000 species identified to date are greater than the combined numbers for the other classes [1].

This taxonomic diversity is reflected in a diversity of body forms, lifestyles, and physiologies, which also reflect the great diversity of aquatic systems that fish inhabits, from freshwater to hypersaline waters, with temperatures ranging from below freezing to >45°C, pressures ranging from 1 to 1000 Atm, and other variability's in solar radiation, oxygen concentrations, ionic and organic matter compositions, turbulence, bottom environments, and so forth. This diversity has driven the use of fishes for scientific inquiries into the influences of environmental variables on the evolution and adaptations of organisms [2]. The study of mechanisms of toxicity in fish involves two distinct rationales, with fish serving both as targets and as models. Clearly, fish are targets for environmental contaminants; some fish populations are highly exposed to chemicals, especially in urban, industrial, or coastal sites (e.g. Yellowfin seabream) [3]. These fish are studied because of concern for population-level effects or as sentinels for the health of the environment. In addition, they may be subjects of research to investigate the mechanisms involved in physiological adaptation to **chemical exposure** [4].

Fish also are widely used as animal models in toxicological research, with the ultimate goal of extrapolating the results to inform questions concerning the potential human health effects of chemical exposure. Thus, research in fish toxicology involves a two-way extrapolation, with human results being used to establish mechanisms of relevance to fish as targets, and fish being used as models to address specific mechanistic questions with application to human health.

Several criteria should be used in selecting **fish species** for ecotoxicological testing [2]:

- Widely available fish species with adequate background information.
- Fish species with a known history of diverse sensitivities and behavioral strategies.
- Indigenous, representative fish species of the receiving system being studied should be considered first. If

Citation: Hedayati A (2016) Yellowfin Seabream, A Target Species for Monitoring of Coastal Pollutants. J Coast Zone Manag 19:e116. doi:10.4172/2473-3350.1000e116

definable and possible, a keystone fish species should be used. Keystone species are those that influence a community because they play a significant role in the community, not because of their numerical dominance.

- Species must be amenable to culturing and maintenance in the laboratory.
- Test species can be collected from wild populations, cultured in the laboratory or purchased from a commercial supplier.

YELLOWFIN SEABREAM, SUITABLE MARINE FISH SPECIES

Yellowfin Seabream (*A. latus*) known as Grey Bream, Japanese Silver Bream, Yellow Sea Bream, Yellow Seabream, Yellow-Finned Black Porgy). It is belongs to Phylum: Chordata (chordates), Subphylum: Vertebrata (**vertebrates**) Infraphylum: Gnathostomata (jawed vertebrates), Superclass: Osteichthyes (bony fishes), Class: Actinopterygii (ray-finned fishes), Order: Perciformes (perch-like fishes), Suborder: Percoidei, Family: Sparidae, Genus: Acanthopagrus and Species: Acanthopagrus latus [5] **(Figure 1)** with max size 50.0 cm TL (male/unsexed); and max weight: 1500 g. It lives in demersal environment of freshwater, brackish and **marine water** within 34°N - 20°S, 48°E - 154°E (including Indo-West Pacific: Persian Gulf and along the coast of India to the Philippines, north to Japan, south to Australia). It Occurs in shallow**coastal** waters and enters river mouths and estuaries. Often feeds in tidal flats, mainly on echinoderms, worms, crustaceans and mollusks [6].

Figure 1: Yellowfin seabream distribution and morphology (modified by www.Fishbase.org).

A. latus is a protandrous hermaphroditic sparid (Porgies) [3]. Yellowfin **seabream**, the natural inhabitant of the creeks studied, are found in the shallow water of the intertidal zone that experiences dual fluctuations in both temperature and salinity. This fish is under threat because of their continuous exposure to toxic chemical-rich industrial effluents that are discharged into the creeks.

Yellowfin sea bream is considered to be one of the most commercially important marine fish in Iran, due to its consumer preference and is newly successful cultured in the coastal area of Iran, especially in Khuzestan province. These economically important fish are under threat because of their continuous exposure to toxic chemical-rich industrial effluents that are discharged into the Persian Gulf and their enclosed creeks. wild stocks of sea bream have declined substantially in recent years [3], owing to overexploitation, pollution and illegal fishing practices, there is an urgent need to revive them through processes of conservation, management, and **mariculture**, knowledge of the mainly common pollution effect on vital organ of this fish, is a basic prerequisite for successful management, and also for correct interpretation of ecological investigations [7], so yellowfin sea bream, bottom-dwelling fish that bring them in intimate contact with sediment during parts or all of their life, is suitable candidate for toxicology testing.

REFERENCES

1. Cossins AR, Crawford DL (2005) Fish as models for environmental genomics. Nature Rev Genet 6: 324-333.
2. Munawar M (2005) Ecotoxicological Testing of Marine and Freshwater Ecosystems. Taylor & Francis Group, LLC: 293.
3. Hesp SA, Potter IC, Hall NG (2004) Reproduction biology and protandrous hermaphroditism in A. latus. Environ Biol Fish 70: 252-272.
4. Hedayati A, Hosseini AR (2012) Endocrine Disruptions Induced by artificial induction of Mercury on Seabream. Comparative Clinical Pathology.
5. Abol-Munafi AB, Umeda S (1994) The gonadal cycle of the yellowfin porgy, A. latus (Houttuyn) reared in the net cage at Tosa Bay, Japan. Suisanzoshoku 42: 135-144.
6. Fishbase 2010. www.fishbase.org
7. Hedayati AS, Savari A, Movahedinia A, Zare P, Bagheri T (2011) Serum Biochemical Change Induced by In vitro Sub Chronic Mercury Chloride in Yellowfin Sea Bream. Iranica Journal of Energy & Environment 2: 208-214.

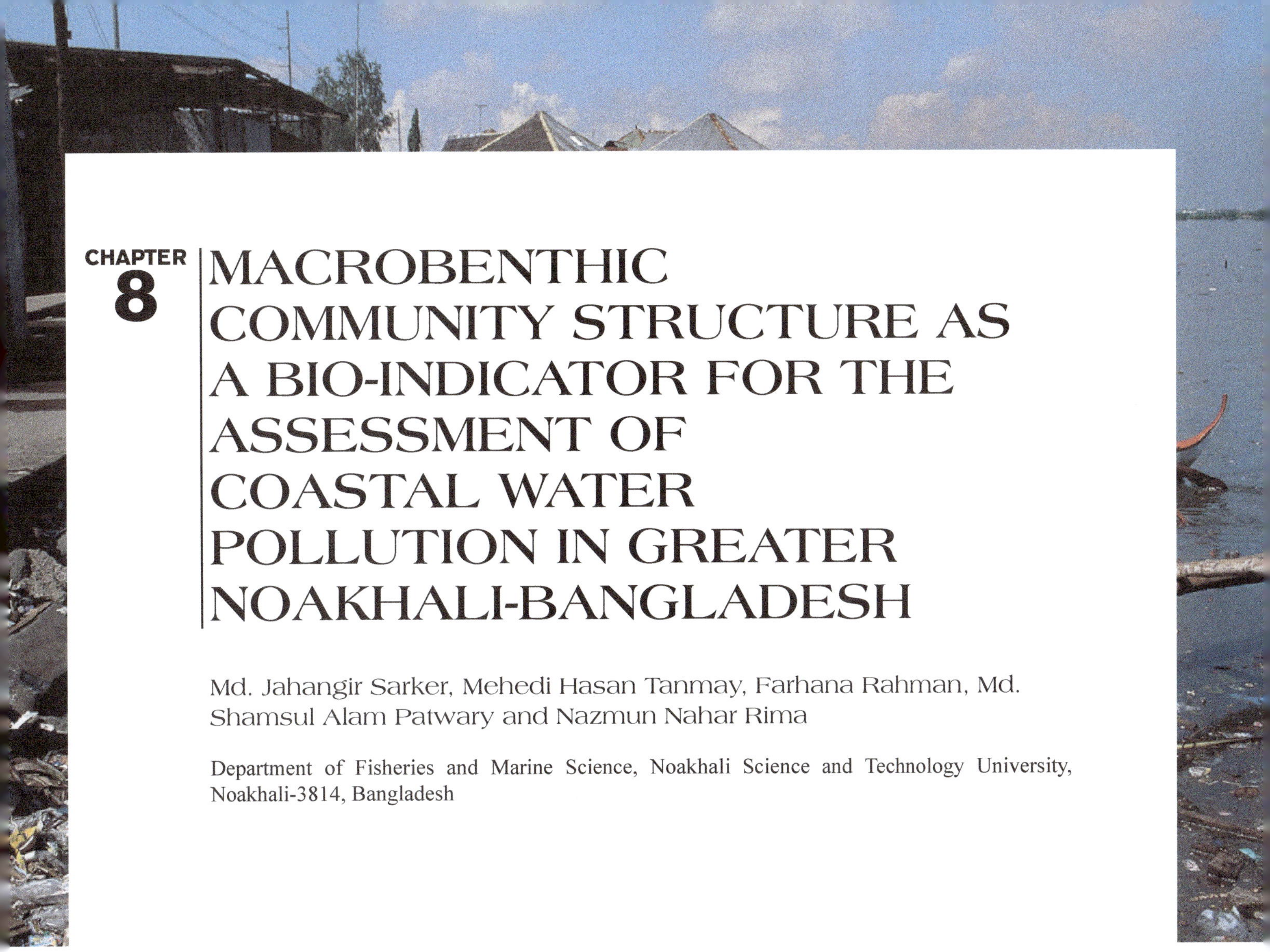

CHAPTER 8

MACROBENTHIC COMMUNITY STRUCTURE AS A BIO-INDICATOR FOR THE ASSESSMENT OF COASTAL WATER POLLUTION IN GREATER NOAKHALI-BANGLADESH

Md. Jahangir Sarker, Mehedi Hasan Tanmay, Farhana Rahman, Md. Shamsul Alam Patwary and Nazmun Nahar Rima

Department of Fisheries and Marine Science, Noakhali Science and Technology University, Noakhali-3814, Bangladesh

INTRODUCTION

Macrobenthos are commonly used as bio-monitors to detect pollution impacts in estuaries [1] for their exclusive and unique characteristics. They are abundant, easy to collect and very diverse with representatives from many different phyla [2] utilizing many different habitats and feeding strategies [3-6]. These assemblages respond predictably to pollution [7,8], integrates the stress over months to years and relatively sedentary in nature [5,9]. Besides this, **macrobenthic** communities are eaten by other higher tropical organisms like fin and shell fishes as food and recycle the organic matters and debris like an ecological engineer [10]. They also provide a linkage between substratum and sea bed and water column predators [11]. Macrobenthos are an important role playing community in aquatic ecosystem because they mineralize, promote and mix the oxygen flux into the sediment which recycle the organic matter [12]. Benthic community determines the amount of nutrients release of the sediments [13]. Biological interactions, such as predation and competition affect the macrobenthic community structure by acting on recruitment, survival, or migration of organisms [14-18]. The characteristics of the life cycle of the species and the influence of temporal fluctuation of abiotic factors, such as environmental temperature or salinity can also change macrobenthic communities in a cyclic pattern over time [19-22]. Physical and chemical factors affect the distribution and abundance of macrobenthic community like contaminations of sediments environment, current of the water organic contents of the sediments, depth, rapid sedimentation and toxicity of sediments causes shifts of macrobenthic communities towards lower abundance [23].

Although much studies on macrobenthic community has been carried out globally however there is scanty published documents on macrobenthic community of coastal area in **Bangladesh**. There are few works on Hatia and Nijhum Dweep Island but the attached part of Hatia with main land of Noakhali district remain untouched. So this study was aimed for the documentation of the abundance and composition of microbenthic community of the branch of lower part of the mighty Meghna River. Therefore the objective of the present study was to know the monthly abundance of macrobenthic community, to know the species composition and diversity of macrobenthic communities in estuarine waters and to assess the status of pollution in Chairman Ghat, Noakhali.

MATERIALS AND METHODS

Sediment and water samples were collected from the intertidal zone of the estuary of a branch of the Meghna River at Chairman Ghat during January, February and March, 2015 from 3

Citation: Sarker J, Tanmay MH, Rahman F, Patwary MSA, Rima NN (2016) Assessment of Coastal Water Pollution In Greater Noakhali-Bangladesh. J Coast Zone Manag 19:427. doi:10.4172/2473-3350.1000427

sampling stations namely S1 (Station 1) (22° 30' 48.3876" N, 91°5' 6.6078" E), S2 (Station 2) (22° 31' 5.2278" N, 91° 5' 26.4768" E) and S3 (Station 3) (22° 31' 34.1868" N, 91° 5' 41.5566" E). An Ekman dredge (mouth opening of 0.0225 m^2) was used to collect sediment samples with triplicate fashion from each station. Subsurface water samples were collected from three sites during high tide condition for measuring water temperature (0C), salinity (ppt), pH, DO (**dissolved oxygen**, ppm) and transparency (cm). Temperature (0C) was measured using a centigrade thermometer in study area where water salinity, pH, transparency and DO were measured*in situ* by using **Refractometer** (INDEX, Model No. REF 201), Digital pen pH meter (HANNA Instrument, Model No. H196107), Secchi disc (20 cm diameter) and DO meter (LUTRON, Model No. DO-5509) respectively.

Sampled sediments were sieved (through 500 μm mesh) to retain macrobenthos and then preserved immediately in the plastic container with other residues in 10% formalin solution. It was then labeled and for further analysis transferred to the laboratory. Small amount of "Rose Bengal" was added to make the macro organism visible. Benthic macrofauna was sorted manually putting on a tray under sufficient light and enumerated up to major taxa. Magnifying glass and microscope (Model No. XSZ 21-05DN, China) often were used for the identification of macrobenthos. Abundance of macrobenthic fauna was calculated in individual per m^2. Sample was then preserved in 70% alcohol (Ethanol). Statistical analysis was performed using SPSS (Version 20) and PAST [24]. Occurrence of macroinvertebrates occurrence (N) was calculated using the following formula [25],

$$N = \frac{O}{a.s} \times 10000$$

Here,

s = Number of sample.

a = Mouth opening area of Ekman dredge in cm^2, and

O = Actual counted number of macro-invertebrates per sampled area,

The dominance (D) index [26] was determined by using following formula:

$$D = \sum_{i-1} \left(\frac{ni}{n}\right)^2$$

Here,

n = Total individual number.

ni = Individuals of species number *i*.

The Simpson Index (1-D) was measured by the following formula [27],

$$(1\text{-}D) = \frac{1}{\sum_{i-1}^{S} Pi\ 2}$$

Here,

P = Proportion (n/N) of individuals of one particular species found (n) divided by the total number of individuals found (N),

S = Number of species.

Shannon-Wiener Index (Species diversity, H') were calculated according to Wilhm and Dorris [28] formula,

$$H' = -\sum_{i-1}^{S} Pi\ Ln\ Pi$$

Here,

Pi = *ni/N* = Proportion of individuals of the total sample of the ith species.

N = Individual number of all the species in total,

ni = Individuals number of the ith species, and

S = Number of species.

The Margalef 's Index (D') was calculated by the following index [29],

$$D' = \frac{S-1}{\ln N}$$

Here,

N = Individuals number in total of sample

S = Species number in sample

ln = Normal log

Equitability (J) was measured by using the following formula [26],

$$J = \frac{H'}{\text{Log } S}$$

S = Species number in a population.

H' = Shannon and Weiner Index

RESULTS AND DISCUSSION

The measured water quality parameter is recorded in **Table 1**. The highest water temperature was recorded (320C) in station S3 in March with mean of 30 ± 0.96 while the lower was observed in January (22.33 ± 0.47). Higher **salinity** was measured in February (10.33 ± 1.25) being the highest at station S3 (12 ppt) comparing to other two months while DO was lowest in this month (4.0 ± 0.29 ppm) in relation to January (6.8 ± 0.24 ppm) and March (5.3 ± 0.73 ppm). PH showed no major variation in January (7.67 ± 0.31), February (7.77 ± 0.21) or in March (7.28 ± 0.70) while transparency was recorded very low in all 3 month of the present study area in January (3.17 ± 1.03 cm), February (4.17 ± 0.85 cm), and in March (1.5 ± 0.71cm). All the parameters varied with the seasonal variation accordingly. Salinity and transparency may fluctuated due to the water availability in different month from January to March because usually March is more close to the monsoon season where January falls under Winter season in this area. A Pearson's correlation was run to determine the relationship between these 5 measured parameters (temperature, salinity, DO, pH and **transparency**) from three stations (S1, S2 and S3) along with abundance of **macrofauna** found during January, February and March, 2015 from the present study (**Table 2**). There was a very strong negative correlation between salinity and DO (r = - 0.859, p >0.01).

Table 1: Water quality parameter measured from three stations during January, February and March (2015).

Month	Station	Temperature (°C)	Salinity (ppt)	Dissolve Oxygen (ppm)	pH	Transparency (cm)
January	S1	22	5	7.1	7.5	4.5
	S2	22	4	6.5	7.4	3
	S3	23	6	6.8	8.1	2
	Mean ± SD	22.33 ± 0.47	5.00 ± 0.82	6.8 ± 0.24	7.67 ± 0.31	3.17 ± 1.03
February	S1	27	9	3.6	7.8	3
	S2	27	10	4.3	7.5	5
	S3	28	12	4.1	8	4.5
	Mean ± SD	27.33 ± 0.47	10.33 ± 1.25	4.0 ± 0.29	7.77 ± 0.21	4.17 ± 0.85
March	S1	30	5	6.2	7.28	3
	S2	31	6	6.8	7.55	2
	S3	32	8	5.3	8.75	1.5
	Mean ± SD	30 ± 0.96	5 ± 1.41	5.3 ± 0.73	7.28 ± 0.70	1.5 ± 0.71

A total number of 14 taxa (families) of macrofauna under 4 class identified from January to March yielded 5481 ind./m^2 with a mean density of 609 ± 445.32 ind./m^2 from all stations (**Table 3**). The highest density of macrobenthic animal was found 919 ind./m^2 in station S1 during March while the lowest was 326 ind./m^2 at station S3 during February (**Table 3**). Lumbrineridae almost dominated each months in all stations contributing 49% (**Figure 1**) of the total macrofauna while Goniadidae and Nereididae were also abundant in all months with a mean of (296 ± 139.33 ind./m^2), (71 ± 42.35 ind./m^2) and (127 ± 88.37 ind./m^2) with a total of (2667 ind./m^2), (637 ind./m^2) and (1141 ind./m^2) respectively (**Figure 2**). In regard of the rest 11 families, their attendance were scattered in different months at different stations. Nephtyidae was found only in March while Terebridae was found in January only at station S3 (15 ind./m^2) and Oedicerotidae was only at station S1 (44 ind./m^2) in March. Variation of microbenthic density might be due to the seasonal variations of environmental parameters [30]. Observed macrobenthic diversity indices were positively correlated with salinity and pH (p <0.05) and negatively correlated with clay and organic matter [31]. Although species level identification of microbenthic animal provides better understanding of environmental impact assessment [32], present results with the family of microbenthic individual is also acceptable [33].

Quantitative distribution of intertidal macrobenthic community structure of the Chairman Ghat has been furnished in **Table 4**. The fauna comprised a total 4 groups including Polychaete, Gastropoda, Bivalvia and **Arthropods** where Polychaete was dominant in all the stations occupying the highest percentage (**Figure 3**) in all three months during all the study period that yield a total (4815 ind./m^2) with a mean value of (535 ± 175.63 ind./m^2) comprising 87.85% of total abundance. Gastropoda was found only at station S3 (15 ind./m^2) and Bivalvia at station S1 (15 ind./m^2) and at S3 (30 ind./m^2) in January. Arthropods were common in all month after the polychaete where polychaete was 83% and Arthropods was 7% in January, polychaete was 96% and Arthropods was 1% in February and polychaete was 84% and Arthropods was 8% in March (**Figure 3**). In all the comparative study of benthic group it is clear that the Polychete is the most abundant and dominant group in all three month. The density of macrobenthos group found in 3 station during 3 months of study period from Chairman Ghat are tabulated in **Table 5** that shows the abundance in individual per square meter (ind./m^2), their station wise percentage and the rank of abundance in 3 months. It shows that Polychete is ranked in number 1 in all stations while other groups fluctuated within different stations in different months where sometime some groups were totally absent. Similar findings was reported near the study area [34] where Polychaeta was dominant [10,35].

Diversity indices

Table 2: Pearson correlation (r) among the measured parameters during the present study period.

Pearson Correlation (r)	Temperature	Salinity	DO	pH	Transparency	Abundance
Temperature	-	0.356	-0.37	0.3	-0.349	0.301
Salinity	-	-	-0.859**	0.4	0.398	-0.207
DO	-	-	-	0	-0.265	-0.141
pH	-	-	-	-	-0.472	-0.293
Transparency	-	-	-	-	-	-0.01
Abundance	-	-	-	-	-	-

Table 3: Abundance of macrobenthic family (ind./m2) at all stations during the study period.

Month	January			February			March						
Station	**JS1**	**JS2**	**JS3**	**FS1**	**FS2**	**FS3**	**MS1**	**MS2**	**MS3**	**Mean±SD**	**Total**	**Percentage**	**Rank**
Lumbrineridae	148	207	222	563	504	193	370	222	237	296 ± 139.33	2667	48.65	1
Goniadidae	59	30	15	119	148	30	59	74	104	71 ± 42.35	637	11.62	3
Nephtyidae	0	0	0	0	0	0	104	30	15	16 ± 32.34	148	2.7	5
Nereididae	281	193	74	74	30	0	163	104	222	127 ± 88.37	1141	20.81	2
Polynoidae	30	15	0	0	0	0	0	0	0	5 ± 9.88	44	0.81	12
Syllidae	0	0	0	0	0	0	0	0	30	3 ± 9.31	30	0.54	13
Sabellidae	0	0	0	0	15	0	0	0	0	2 ± 4.66	15	0.27	15
Capitellidae	15	0	0	30	0	44	44	0	0	15 ± 18.48	133	2.43	7
Terebridae	0	0	15	0	0	0	0	0	0	2 ± 4.66	15	0.27	14
Isaeidae	15	44	0	0	0	0	0	0	0	7 ± 14.16	59	1.08	8
Mytilidae	15	0	30	0	0	0	0	0	0	5 ± 9.88	44	0.81	9
Oedicerotidae	0	0	0	0	0	0	44	0	0	5 ± 13.97	44	0.81	10
Mysidae	0	30	0	0	0	15	0	0	0	5 ± 9.88	44	0.81	11
Portunidae	30	0	0	0	0	0	89	30	0	16 ± 28.32	148	2.7	6
Unidentified	30	15	44	15	0	44	44	59	59	35 ± 19.75	311	5.68	4
Total	622	533	400	800	696	326	919	519	667	609 ± 445.32	5481	100	
Family Found	8	6	5	4	4	4	7	5	5		14		

JS1, JS2 and JS3 = S1, S2 and S3 in January respectively; FS1, FS2 and FS3 = S1, S2 and S3 in February respectively; MS1, MS2 and MS3 = S1, S2 and S3 in March respectively.

Diversity Indices of different station characterized the present study (**Table 6**). Different diversity indices showed significant difference between the months (**Figure 4**). Higher values of the diversity indices were observed in January and March while in February the values were lower than other two months. The highest value for Dominance (D) was in February (0.5438). The Simpson (1-D) value was highest (0.728) in March and lowest (0.4562) in February. The Shannon (H′) diversity index is another important one. In the present investigation it ranged from 0.7797 at station S2 in February to 1.648 at station S1 in March where 1.514, 0.9202 and 1.571 was for January, February and March respectively. The Margalef's index of benthic macro-invertebrates at station S1 (0.4501) was lowest maintaining the value 0.6689 in February for in total and at station S1 (1.096) was highest maintaining the value 1.234 in January for in total while that value was 0.9246 in March. While the value of Equitability (J) index of benthic macro-invertebrates at station S2 (0.5452) was lowest maintaining the value 0.4183 in February for in total and at station S2 (0.7602) was highest maintaining the value 0.6016 in March for in total while that value was 0.4546 in January respectively.

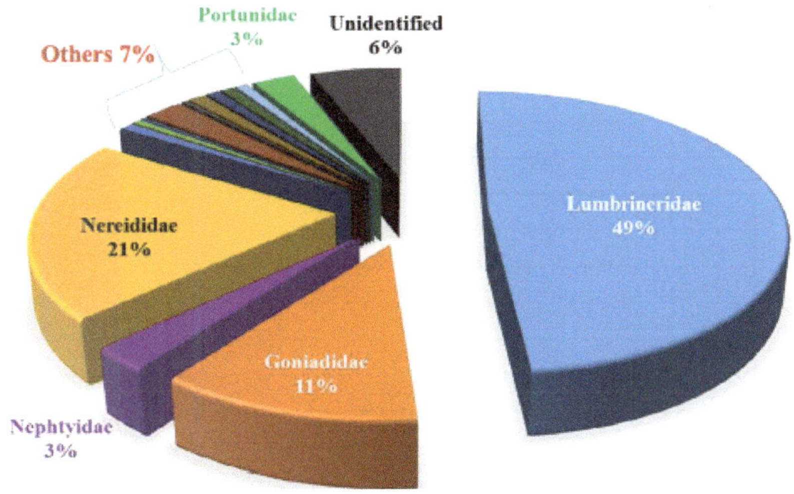

Figure 1: Composition of macrobenthic animals (family) observed in the present study.

Shannon-Weaver diversity index was defined [36] as H′ and the highest and lowest value of H′ indicated rich diversity with healthier environment and poor diversity with polluted environment respectively. Water bodies with macro-benthos Shannon-Wiener diversity index <1 are classified as heavily polluted, 1-3 is moderately polluted and >3 clean is environment [28,37-40]. So it can be said that all the selected stations in the study area was heavily polluted in February (H′ = 0.9202) while pollution was moderate in January (H′ = 1.514) and March (H′ = 1.571). In one hand the suitability of habitat is reflected by the higher diversity values for the organism while it is reported that high species diversity is correlated with longer food chain, complex food web and more stable community on the other hand [41]. Depending on the number of species **Margalef index** (D′) shows variation having no limit on tis value. So it can be postulated that the present study area is moderately polluted according the results calculated from Margalef index which is compatible with the findings of other research [34].

Table 4: Group wise total abundance distribution of macrobenthos in all stations in Chairman Ghat observed during the present study.

Month	January			February			March					
Station	**JS1**	**JS2**	**JS3**	**FS1**	**FS2**	**FS3**	**MS1**	**MS2**	**MS3**	**Mean ± SD**	**Total**	**Rank**
Polychaete	533	444	311	785	696	267	741	430	607	535 ± 175.63	4815	1
Gastropoda	0	0	15	0	0	0	0	0	0	2 ± 4.66	15	5
Bivlvia	15	0	30	0	0	0	0	0	0	5 ± 9.88	44	4
Arthropods	44	74	0	0	0	15	133	30	0	33 ± 42.92	296	3
Unidentified	30	15	44	15	0	44	44	59	59	35 ± 19.75	311	2
Total	622	533	400	800	696	326	919	519	667	609 ± 176.74	5481	

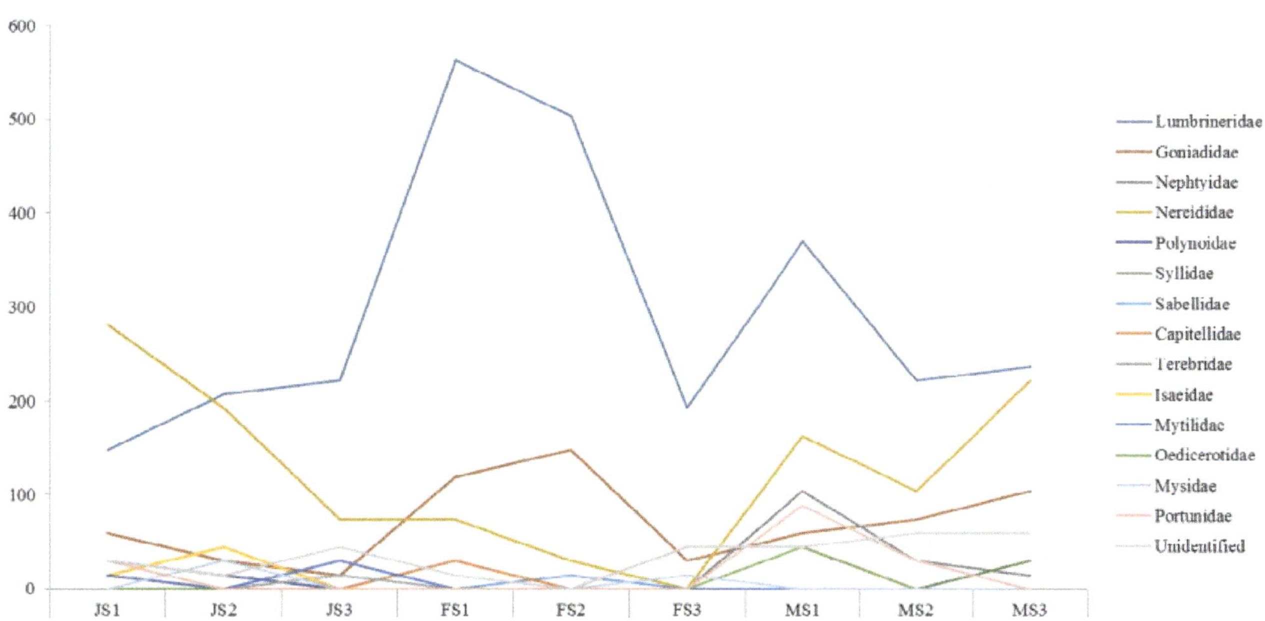

Figure 2: Monthly variation in abundance of macrobenthic animal's family (ind./m2) in three station

JS1, JS2 and JS3 = S1, S2 and S3 in January respectively; FS1, FS2 and FS3 = S1, S2 and S3 in February respectively; MS1, MS2 and MS3 = S1, S2 and S3 in March respectively.

Table 5a: Density and station wise rank of macrobenthic groups based on percentage found during January of the study period

Station	Groups	Ind./m2	Percentage (%)	Station wise Rank
JS1	Polychaete	533	85.74	1
	Gastropoda	0	0	5
	Bivlvia	15	2.38	4
	Arthropods	44	7.15	2
	Unidentified	30	4.76	3
	Total	622	100	
JS2	Polychaete	444	83.39	1
	Gastropoda	0	0	4
	Bivlvia	0	0	5
	Arthropods	74	13.9	2
	Unidentified	15	2.78	3
	Total	533	100	
JS3	Polychaete	311	77.78	1
	Gastropoda	15	3.7	4
	Bivlvia	30	7.41	3
	Arthropods	0	0	5
	Unidentified	44	11.11	2
	Total	400	100	

JS1, JS2 and JS3 = S1, S2 and S3 in January respectively.

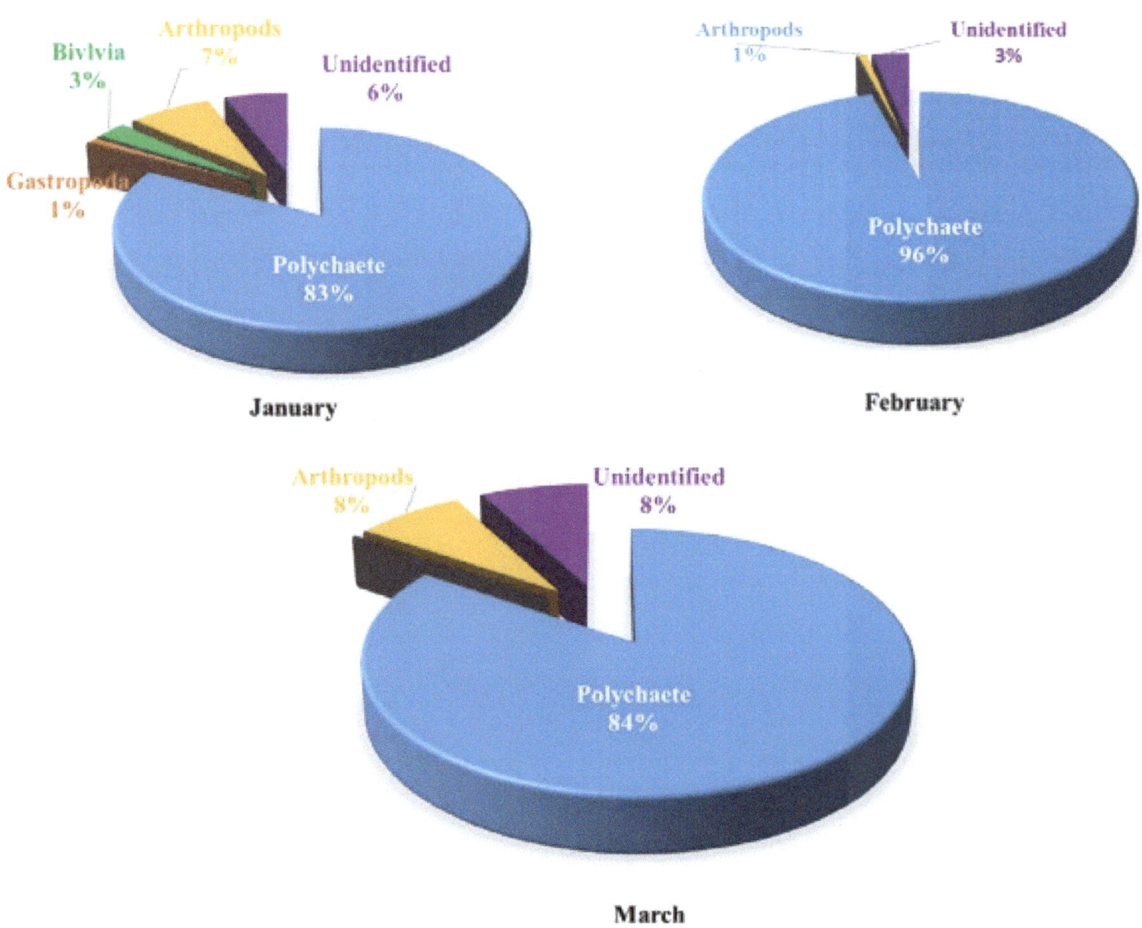

Figure 3: Composition of macrobenthic animal (group) observed during the study period (J- in January, F- in February and M- in March).

Table 5b: Density and station wise rank of macrobenthic groups based on percentage found during February of the study period.

Station	Groups	Ind./m2	Percentage (%)	Station wise Rank
FS1	Polychaete	785	98.15	1
	Gastropoda	0	0	3
	Bivlvia	0	0	4
	Arthropods	0	0	5
	Unidentified	15	1.85	2
	Total	800	100	
FS2	Polychaete	696	100	1
	Gastropoda	0	0	2
	Bivlvia	0	0	3
	Arthropods	0	0	4
	Unidentified	0	0	5
	Total	696	100	
FS3	Polychaete	267	81.8	1
	Gastropoda	0	0	4
	Bivlvia	0	0	5
	Arthropods	15	4.54	3
	Unidentified	44	13.63	2
	Total	326	100	

FS1, FS2 and FS3 = S1, S2 and S3 in February respectively.

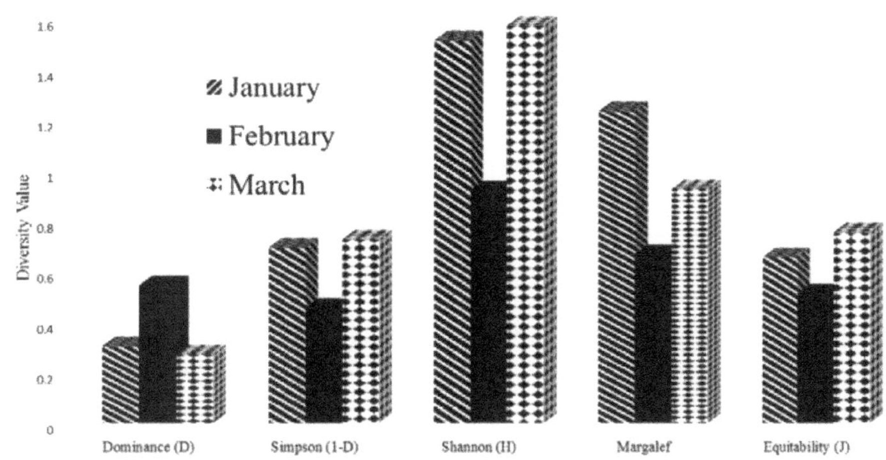

Figure 4: Monthly Values of different diversity indices of the present study.

Table 5c: Density and station wise rank of macrobenthic groups based on percentage found during March of the study period

Station	Groups	Ind./m^2	Percentage (%)	Station wise Rank
MS1	Polychaete	741	80.6	1
	Gastropoda	0	0	4
	Bivlvia	0	0	5
	Arthropods	133	14.51	2
	Unidentified	44	4.84	3
	Total	**919**	**100**	

MS2	Polychaete	430	82.78	1
	Gastropoda	0	0	4
	Bivlvia	0	0	5
	Arthropods	30	5.71	3
	Unidentified	59	11.42	2
	Total	**519**	**100**	
MS3	Polychaete	607	91.07	1
	Gastropoda	0	0	3
	Bivlvia	0	0	4
	Arthropods	0	0	5
	Unidentified	59	8.88	2
	Total	**667**	**100**	

MS1, MS2 and MS3 = S1, S2 and S3 in March respectively.

Table 6: Different diversity Indices at different stations observed during the study period.

Month	Station	Taxa (S)	Individuals	Dominance (D)	Simpson (1-D)	Shannon (H')	Margalef (D')	Equitability (J)
January	**JS1**	8	592.58	0.305	0.695	1.507	1.096	0.7245
	JS2	6	518.51	0.3127	0.6873	1.374	0.7999	0.7666
	JS3	5	355.54	0.4445	0.5555	1.092	0.681	0.6787
	Total	10	1466.65	0.3044	0.6956	1.514	1.234	0.6576
February	**FS1**	4	785.18	0.5472	0.4528	0.8703	0.4501	0.6278
	FS2	4	696.29	0.5709	0.4291	0.7797	0.4583	0.5625
	FS3	4	281.47	0.5069	0.4931	0.943	0.5319	0.6802
	Total	6	1762.95	0.5438	0.4562	0.9202	0.6689	0.5136
March	**MS1**	7	874.06	0.2485	0.7515	1.648	0.8859	0.8468
	MS2	5	459.25	0.3195	0.6805	1.335	0.6526	0.8296
	MS3	5	607.4	0.3183	0.6817	1.275	0.6241	0.7921
	Total	8	1940.74	0.272	0.728	1.571	0.9246	0.7557
All Station		14	5170.35	0.3327	0.6673	1.509	1.52	0.5717

JS1, JS2 and JS3 = S1, S2 and S3 in January respectively; FS1, FS2 and FS3 = S1, S2 and S3 in February respectively; MS1, MS2 and MS3 = S1, S2 and S3 in March respectively.

CONCLUSION

14 families under 4 major groups/taxa yielded a total of 5481 ind./ m^2 with a mean density of 609 ± 445.32 ind./m^2 from all stations. The Shannon-Wiener index indicate that all the stations selected was heavily polluted in February (H' = 0.9202) while pollution was moderate in January (H' = 1.514) and in March (H' = 1.571). The Margalef Species Richness Index (D') value was highest in January (1.234) and lowest in February (0.6689) where March (0.9246) maintained the medium value. Which reviles that the area falls under present study was moderately polluted.

REFERENCES

1. Smith RW, Weisberg SB, Cadien D, Dalkey A, Montagne D, et al. (2001) Benthic response index for assessing infaunal communities on the southern California Coastal Shelf. Ecological Applications 11: 1073-1087.
2. SnelgrovePVR (1998) The biodiversity of macrofaunal organisms in marine sedi-ments. Biodiversity and Conservation 7: 1123-1132.
3. Rhoads DC (1974) Organism-sediment relations on the muddy sea floor. Oceanography and Marine Biology Annual Review 12: 263-300.
4. Fauchald K, Jumars PA (1979)The diet of worms: a study of polychaete feeding guilds. Oceanography and Marine Biology Annual Review 17: 193-284.
5. Weisberg SB, Ranasighe JA, Schaffner LC, Diaz RJ, Dauer DM, et al.(1997) An estuarine index of biological integrity (B-IBI) for Chesapeake Bay. Estuaries 20: 149.
6. Little C (2000) The Biology of Soft Shores and Estuaries. Oxford University Press, New York, NY: 158.
7. Pearson TH, Rosenberg R (1978)Macrobenthic succession in relation to organic enrichment and pollution of the marine environment. Oceanography and Marine Biology Annual Review 16: 229-311.
8. Hart CW, Fuller SLH (1979) Pollution Ecology of

Estuarine Invertebrates. Academic Press, New York: 406.

9. Paul JF, Scott KJ, Campbell DE, Gentile JH, Strobel CS (2001) Developing and applying a benthic index of estuarine condition for the Virginian Biogeographic Province. Ecological Indicators 1: 83-99.
10. Asadujjaman M, BelalHossain M, Shamsuddin M, Amin MA, Azam AKM(2012)Occurrence and Abundance of Macrobenthos of Hatiya and NijhumDweep Islands, Bangladesh. Middle-East Journal of Scientific Research 11: 184-188.
11. Gray JS, Elliott M(2009) The Ecology of Marine Sediments (2ndedn). The Oxford University Press, Oxford, UK: 256.
12. Lind OT (1979) Handbook of common method in limnology (2ndedn). The C.V. Mosby Company, St. Louis: 136-145.
13. Newrkla P, GunatilakaA (1982)Benthic community metabolism of three Austrian pre-alpine lakes of different tropic conditions and its oxygen dependency. Hydrobiologia92: 531-536.
14. Woodin SA (1974)Polychaete abundance patterns in a marine soft-sediment environment: the importance of biological interactions. Ecological Monographs 44: 171-187.
15. Peterson CH, Andre SV (1980) An experimental analysis of interespecific competition among marine filter feeders in a soft-sediment environment. Ecology 61: 129-139.
16. Brenchley GA (1982) Mechanisms of spatial competition in marine soft-bottom communities. Journal of Experimental Marine Biology and Ecology 60: 17-33.
17. Thrush SF, Pridmore RD, Hewitt JE, Cummings VJ (1992) Adult infauna as facilitators of colonization on intertidal sandflats. Journal of Experimental Marine Biology and Ecology 159: 253-265.
18. Knox GA (2000) The ecology of sea shores. CRC Press, New York: 555.
19. Day JW, Hall CAS, Kemp WM,Yañez-Arancibia A (1989)The estuarine bottom and benthic subsystem. Estuarine Ecology. Wiley & Sons, New York: 338-376.
20. Souza JBR, Gianuca NM (1995) Zonation and seasonal variation of the intertidal macrofauna on a sandy beach of Parana State, Brazil. Scientia Marina 59: 103-111.
21. Veloso VG, Cardoso RS, Fonseca DB (1997) Spatio-temporal characterization of intertidal macrofauna at Prainha beach (Rio de Janeiro State). OecologiaBrasiliensis 435: 213-225.
22. Das Neves LP, da Silva P de SR, Bemvenuti CE (2008) Temporal variability of benthic macrofauna on Cassino beach, southernmost Brazil. Iheringia, SerieZoologia 98: 36-44.
23. Pearson TH (1970) The benthic ecology of Loch Linnhe and Loch Eil, a Sea–Loch system on the west coast of Scotland .The physical environment and distribution of the macrobenthic fauna. JExp Mar BiolEcol5: 1-34.
24. Ryan PD, Harper DAT, Whalley JS(1995)Palstat, Statistics for palaeontologists. Chapman & Hall (now Kluwer Academic Publishers).
25. Welch PS (1948) Limnology II edition. Mcgraw Hill book Company, New York.
26. (1999) In:Harper DAT (ed.), Numerical Palaeobiology. John Wiley & Sons.
27. Simpson EH (1949) Measurement of diversity. Nature 163:688.
28. Wilhm JL, Dorris TC(1966) Species diversity of benthic macroinvertebrates in a stream receiving domestic and oil refinery effluents. Am. Midland Nat 76: 427-449.
29. Margalef R (1968) Perspectives in Ecological Theory. University of Chicago Press, Chicago, IL: 111.
30. Islam Shafiqul M,SikdarNurulAzim M, Al-Imran M,HossainBelal M,MallickDebbroto(2013) Intertidal Macrobenthic Fauna of the Karnafuli Estuary: Relations with Environmental Variables, World Applied Sciences Journal21: 1366-1373.
31. HossainBelalM, Marshall David J (2014)Benthicinfaunal community structuring in an acidified tropical estuarine system. Aquatic Biosystems10:11
32. Narayanaswamy BE, Nickell TD, Gage JD (2003) Appropriate levels of taxonomic discrimination in deep-sea studies: species vs family. Marine ecology. Progress series 257: 59-68.
33. Warwick RM (1988) Analysis of community attributes of the macrobenthos of Frierfjord/ Langesundfjord at taxonomic levels higher than species. Marine Ecology Progress Series 46: 167-170.
34. SarkerJMd, PatwarySAMd, Uddin AMMB, Hasan M Md, Tanmay MH, et al. (2016) Macrobenthic Community Structure - An Approach to Assess Coastal Water Pollution in Bangladesh. Fish Aquac J 7: 157.
35. HossainBelal M, Amin AMN, AsadujjamanM, RahmanS (2013)Analyses of Macrobenthos of Hatiya and NijliumDweep Islands at Higher Taxonomic Resolution. Journal of Fisheries and Aquatic Science 8: 526-534.
36. Wilhm JL (1975)Biological indicators of pollution, In: River ecology, WhittonBA (ed.), Studies in Ecology, Blackwell SciPubl, London 2: 375-402.
37. Islam MS (2008) Phytoplanktonic diversity index with reference to MucalindaSerovar, Bodh-Gaya. Proceedings of the Taal' 2007: The 12th World Lack Conference: 462-463.
38. Tanimu Y, Bako SP, AdakoleJA, Tanimu J (2011) Phytoplankton as bioindicators of water quality in saminaka reservoir, Northern Nigeria. Proceedings of the International Symposium on Environmental Science and Technology, Dongguan, Guangdong Province, China: 318-322.
39. Stub R, Appling JW,Hatstetter AM, Hass IJ (1970) The effect of industrial waste of Memphis and Shelby country on primary planktonic producers.Bioscience 20:905-912.
40. Mason CF (1988) Biology of Fresh Water Pollution. Longman scientific and technical.
41. Margalef R (1956) Information Y diversidadespecificaenlascommunidades de organisms. InvestPesg 3: 99-106.

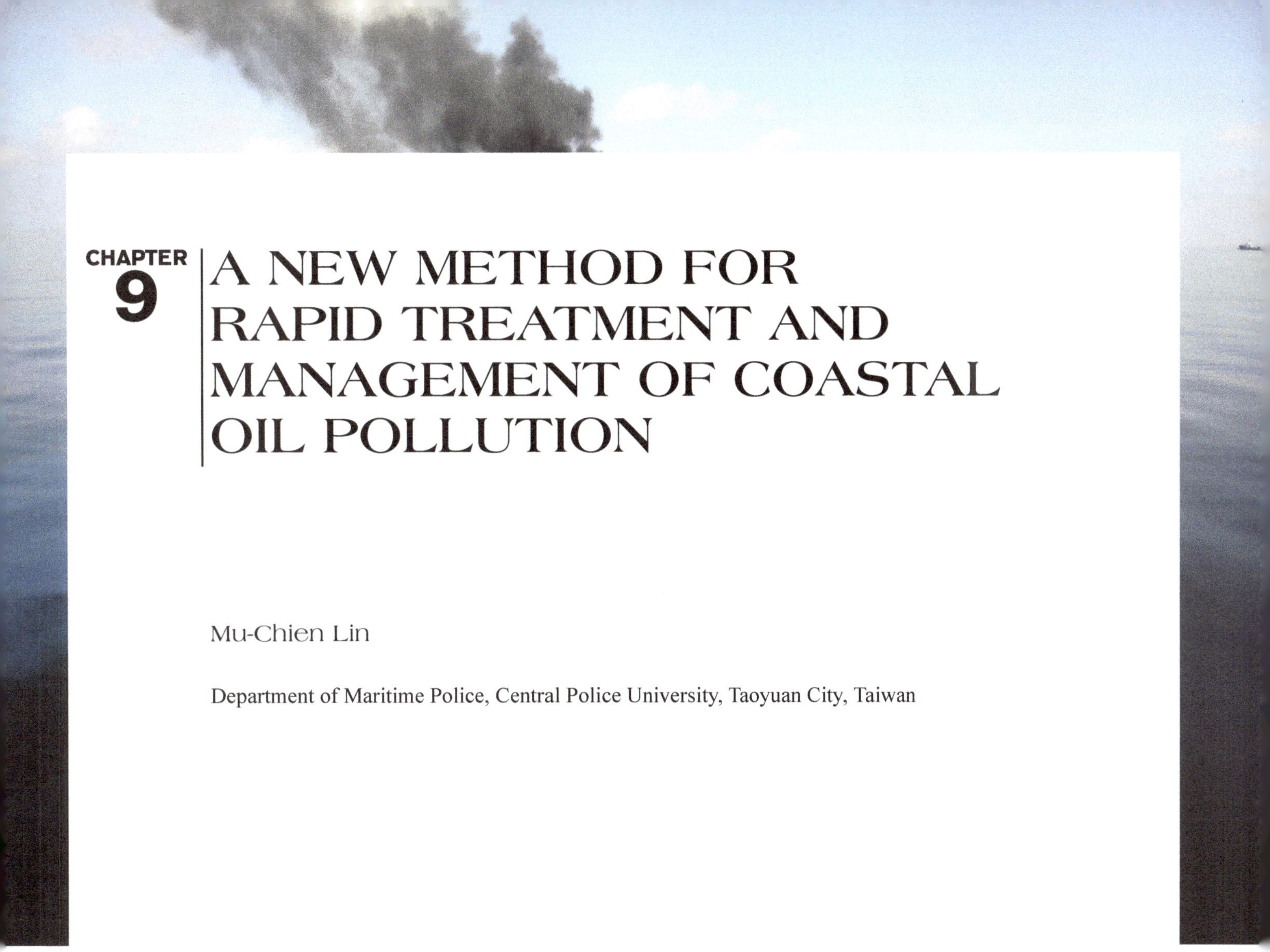

CHAPTER 9

A NEW METHOD FOR RAPID TREATMENT AND MANAGEMENT OF COASTAL OIL POLLUTION

Mu-Chien Lin

Department of Maritime Police, Central Police University, Taoyuan City, Taiwan

INTRODUCTION

Marine accidents occur continuously throughout the oceans of the world. According to the Taiwan Ministry of Transportation, more than 800 accidents occur around Taiwan coastal waters every year, even though Taiwan's waters consist of a small portion of the world's oceans. Marine accidents always lead to oil pollution, which impair our health, prosperity and livelihoods [1]. Hence, the whole world is affected by this crisis of marine pollution. Governments must take pains to prevent oil pollution and adopt methods to swiftly and thoroughly remediate such pollution when it does occur. Etkin et al. [2] showed that the cost of preventing oil pollution is still high, and responses to spills of heavy fuels are more than ten times as expensive as spill responses to lighter crudes and diesel fuels. Most heavy oil pollution incidents were caused by cargo ship or tanker accidents resulting in oil spills, and therefore an efficient method for cleaning up oil pollution that dramatically decreases the costs and eliminates secondary pollution is vitally needed.

Marine accidents occur off the coast and sometimes near **coastlines**. The accidents occurring near coastlines give rise to some of the most serious pollution crises. The effects vary from coast to coast since there are many types of coastline, such as beach, rock, coral reef, fish culture field, mangroves and so on. However, the majority of coastal oil pollution incidents may be beach pollution or coastal rock pollution, although other types of seashore pollution have occurred as well.

With respect to beach pollution, the method employed was to spray the polluted beach with high temperature steam in order to melt the layer of oil, but this caused the oil-water mixture to flow out to sea. Since such mixture still contained oil, it simply became a new source of pollution somewhere else. The job of beach remediation has traditionally been given by governments to subcontractors. But those subcontractors simply collected the pollutants and buried them or washed them away. "Sweeping it under the rug" in this manner may have allowed the oil pollution to be hidden from sight, but the threat to human health and the environment still remained because the oil had not been degraded to safe levels.

Oil pollution on rocky coastlines, moreover, is also a series problem. This is illustrated by in the accident that occurred off Taiwan's coast involving the Greek cargo vessel M/V AMORGOS, which ran aground off the coast of southern Taiwan and caused extremely serious **marine pollution**. Putting aside the direct economic loss, the Taiwan government used more than 35 thousand workers and much material, but the methods employed did not yield good results. The oil stains on the coastal rock was very difficult to clean up and took a long

Citation: Lin MC (2016) A New Method for Rapid Treatment and Management of Coastal Oil Pollution. J Coast Zone Manag 19:424. doi: 10.4172/2473-3350.1000424

time. Essentially, no one could conceive of an efficient method to solve beach and rocky coast pollution problems concurrently. Since the difficulty on treating coastal oil pollution has been a longstanding problem, it has become a matter of great concern. Experiments in biodegradation indicated that to be a cheap, safe and reliable way, gradually making it to prominence as the method currently under research today [3]. Vast amounts of literature have been generated with respect to investigation of biodegradation, resulting in application techniques that improved the viability of this approach [4]. They studied the physical, chemical process as well as effects upon the environmental surroundings and found that biodegradation efficiency depends not only upon the structure of oil pollutant but also upon **environmental factors** [5,6].

Recently, much research has given further consideration in this field. Zahed et al. [7] focused their attention on the study of dispersants. Theoretically, dispersants are a kind of surface-active agent that contains both hydrophobic groups (their tails) and hydrophilic groups (their heads). That is, a surfactant contains both an oil soluble component and a water soluble component, which will diffuse in water and at interfaces between oil and water. In experiments, as the area of contact between oil and water expands, the rate of degradation accelerates. However, in a real-life marine pollution environment, it is unknown whether the microbes stay around the dispersants or not due to the vast expanse of ocean. In reality, combinations of oil and dispersant flow with the tides, ride the waves, sink down into the sea, get ingested by fish and enter the food chain, and kill corals or poison farming areas. The result is dire consequences. Accordingly, in the interests of**environment protection**, it is suggested that the use of dispersants should be avoided whenever possible.

The vast amount of literature devoted to hastening degradation has been reviewed. Gallego et al. [8] attempted degradation of diesel oil by applying fertilizer and found that phosphate or nitrogenous fertilizers are effective to hasten degradation. They focused on the degradation of diesel or **petroleum hydrocarbon** whereas little research focused on the degradation of heavy oil.

Wu et al. [9] developed a method utilizing a system of ponds to treat palm oil mill effluent. Such a centralized system can also be used to treat coastal oil pollution. Unquestionably, a vast area of polluted beach is beyond our control with respect to weather, tides, and destruction by mankind. The idea of a centralized treatment facility eliminates these variables, and optimizes the coastal oil pollution treatment method presented in this paper.

More recently, an application of the nanometer photocatalyst method on decomposition of benzene has been successfully used in the treatment of coastal oil pollution. Lin [10] studied and found that using this method could completely decompose the benzene or gasoline within hours [11]. This initial success found a new path in the treatment of oil pollution [12,13].

The present study is the first attempt to investigate the treatment of coastal oil pollution in a centralized facility. Since the method of**bioremediation** or nanometer photocatalyst used in this study is thoroughly explored elsewhere, it does not fall within the scoped of the present article.

Physical model

Putting aside river shores, seashores are largely divided into beach, rock coast, soil coast, aqua-culture zone, coral reef, or marine leisure areas and so on. Most often, the coastline is mainly beach or rock coast as mentioned in the last section. No matter what kind of coast, however, the polluting oil should be enclosed by an oil boom as soon as possible to prevent the oil from diffusing into the sea. The only disadvantage is that marine oil pollution enclosed by oil boom in the past has gotten out of control due partly to the accident occurring in an adverse climate [14], partly to mutual shirking of responsibility between governments, and partly to delays in negotiations between the government and the culpable shipping company. Therefore, these factors must be mitigated as much as possible. Otherwise, oil pollution will continue to be exacerbated by natural processes or human errors in judgment that will transform otherwise trivial matters into a catastrophe again and again.

Effects of beach oil pollution on ecology

Beaches can be separated into two principal types: those which are formed by erosion or by deposition. The erosion beach, forming a steep beach coast beaten by surf, is composed of large particles that develop a simple ecosystem in which water does not accumulate. The tides are precluded from extending to a large area. For this reason, little serious oil pollution occurs here. Conversely, a deposited beach consists of vast smooth, fine sand on a flat plane so that water percolates into subsoil year after year, and therefore attracts many species (crabs, gulls, microbes, etc.) that symbiotically organize into mini-ecosystems. When oil pollution occurs, layers of oil cover the beach surface, completely cutting off air, water and sunlight. Crabs, gulls and microbes bath together in oil. The chain of symbiosis is cut. The miserable scene, no matter whether macro or micro, is quite hellish. The beach cannot recover unless the oil pollution is completely cleaned up [15].

Failures in treatment on beach oil pollution

As mentioned above, the treatment of beach oil pollution by steam is not acceptable. Apparently, a beach cleared of oil by steam leads to microorganisms being entirely destroyed, making it harder to regenerate than if steam was never used. Additionally, the mixture swept away from the coast becomes a new source of pollution. And the oil that remains never vanishes.

Recently, governments sent excavators to scrape the pollution oil layer off and burry it. But in any event, both the use of steam or excavator scrapping produces the same result. The polluting oil is never degraded.

Currently, bioremediation is directly applied for the treatment of beach oil pollution [16]. Workers enclose the polluted beach to isolate the area and scatter microbes onto the polluted surface. They then apply fertilizer, spray on water and fork the soil, positioned there all day no matter how windy, rainy, cold or hot. Although this may be adequate for small-scale remediation, heavy oil pollution is normally on quite a large-scale when spilled form a ship or oil tanker. On-site bioremediation cannot result in complete oil degradation due to delays and complicated external reasons, such as low temperatures, which hampers microbe metabolism; rain, which dilutes the microbes; wind, which destroys the fences; tides, which washes off the microbes; and salt, which hinders microbial life. Occasionally, even humans treading on the site or animal excrement change the composition of the soil. Although large numbers of workers are employed, they are reluctant to spade the beach under strong sun, or when it's a raining or when it's a cold day. Thus, extending over a period of years, the degradation becomes unsuccessful and a huge amount of time, manpower and money is wasted.

Pollution of coastal rock

To investigate oil adhesion on rock, Lin [14] made a simple study with three stones. The author immersed the first one (stone A) in water, then picked it up and dropped oil on it. When this stone was put into water again, the oil drop appeared to tenuously adhere to the stone because of the incompatibility between oil and water. The oil drop on stone A was removed easily at low water pressure. It shows that oil cannot adhere onto wet stone.

The second stone (stone B) was dried, given a drop of oil and immersed in flowing water. The oil drop on stone B could not be washed away unless high water pressure was used. The viscous force acting on stone B is attributable to the **magnetic force** in the four basic universal forces. The magnetic force between stone surface and oil is larger than between stone surface and water. In order to destroy the larger magnetic force between stone and water by using the smaller magnetic force with water, one must entirely rely on an extra force of impact upon it. The oil drop cannot be removed except through applying a stronger impact. Therefore it is difficult to wash out polluted rocks naturally by the action of sea water.

Furthermore, the third (stone C) was a wet stone, like stone A, which also got a drop of oil, but was exposed under the sun long enough to dry. Stone C grew black because the oil drop remained on the stone's surface while the water evaporated, making it similar to stone B in that water could not wash away the oil drop.

In light of this, we know that oil floats on the sea when oil pollution occurs on coastal rock. The flood and ebb tides alternately work to carry oil over the rocks with the rising tide where it immediately catches onto the rocks' surface at low tide. If the pollutant is cleaned up early enough when the tide is going out, the clean-up would be easier. Conversely, it is difficult to clean up once, due to human delay, the rocks have dried under the sun.

The initial treatment of coastal rock is by manual application of absorbent sheets to clean the thicker layer of oil residue. Even after such cleaning, however, oil remains in polluted coastal rock, producing an unnatural black or brown coloration difficult to remove, which is left for natural degradation to recover over many years.

MATERIALS AND METHODS

The centralized plant method

Governments have simply left **environmental remediation** up to corporations to manage the problems of oil pollution. Such corporations, signed up with government, sent excavators to the polluted site to scrape the pollution oil layer off and transport it to their plant. Nevertheless, most of these corporations were not well set up to deal with the practical exigencies of achieving degradation to within safe levels. Being not able to degrade the pollutant completely, they opted to simply burry the polluted scrapings in order to cut costs. Unfortunately, this caused secondary pollution to the environment. In the final analysis, this faulty policy and ineffectual method led to a detrimental result. It seems feasible that a perfect biodegradation can remedy this flaw.

Recently, some researchers studied the biodegradation of petroleum on beach. But the quality of degradation was not adequate because bad weather or other factors common at polluted sites was hard to control, as mentioned above. Moreover, extreme conditions such as typhoon, hurricane or cold temperatures can make such remediation last for years.

To remedy of these flaws, a simple and quick process can now be employed. The authors combined the ideas of scraping the polluted oil layer with off-site biodegradation and developed a method for a centralized treatment plant. The plant, set up with a multistage washing line, degradation ponds, filtering mechanisms, a furnace and degradation system, can intensively degrade the oily pollutant to safely levels relatively quickly.

The process as below:

- Send excavators to the polluted beach to scrape the polluted oil layer off.
- Transport the oily pollutant to a centralized treatment plant.
- With the oily sand removed, make good and reinstate the now unpolluted beach immediately.
- Wash out the oily pollutant by multistage line, and then carry back the cleaned sand to the beach.
- Filter the oily waste water by multiple layers of oil absorbent sheets.
- Dry the oil absorbent sheets and incinerate them.
- Degrade the residual oil.

The system for a centralized treatment plant is shown as **Figure 1**.

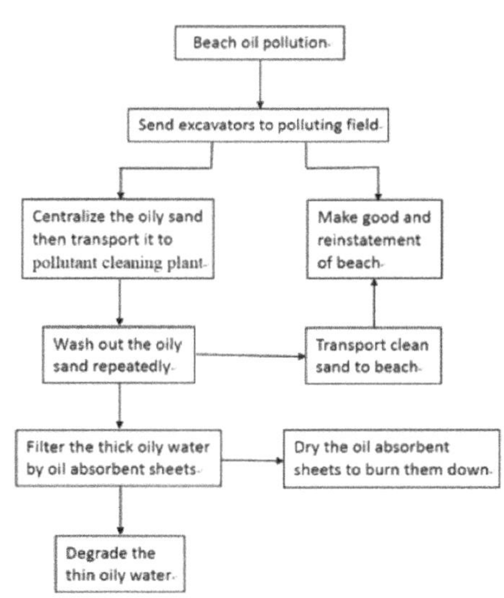

Figure 1: The flow chart of treatment Plant of beach oil pollution.

Most of oil will be separated from pollutants from step (1) to (6), therefore, entirely the residual oil can be degraded in step (7), no matter by biodegradation or photo degradation [11]. Marine accidents usually lead to hundreds of tons of **oil spills**. The Coast Guard should save pollution immediately by oil booms, oil absorbent sheets, skimmers or other equipment. If an oil spills coincide with high tide, most of oil will be removed by Coast Guard and the residual spreading oil is stranded on the beach when the tide ebbs. Although this pollution is not serious, it may cause a large area of pollution. Under the circumstances, the presented method can be used for remediation *in situ*.

METHOD

In 2014, two marine accidents occurred in Penghu and Taitung (Taiwan) causing diesel oil pollution on beach. The pollutants were taken to a centralized plant for each 10 kg. Washing out the oily pollutant twice, the sand was fully washed out, and analyzed in Double-Shot Py-GC system in order to make sure the sand was clean or not. So return the clean sand to beach. The washed oily water was filtered by multiple layers of oil absorbent sheets. Then dry the oil absorbent sheets and incinerate them. By this way most the oil was successfully separate out from pollutants. Finally the tiny oil floating on water was photo degraded by TiO_2 nanometer photo catalysis [11].The photo degradation system used a light with ultraviolet wavelength of 360 nm, which illuminate the photocatalyst to produce peroxide ions. Residual diesel oil was completely decomposed and became CO_2 and H_2O.

Results

The pollutants, presented at the beaches in Taitung (sample 1) and Penghu (sample 2), originally containing diesel oil at 56,000 ppm and 100,000 respectively. Undergo a process of washing out and photo degradation [11]. The residual concentration was only at 56 ppm and 600 ppm respectively. **Table 1** shows the results of present method.

Table 1: The results of treatment.

Sample	Original Concentration (ppm)	Final Concentration (ppm)	Removed oil (%)
1	56,000	56	99.9
2	100,000	600	99.4

The washing time was 1 day, whereas the photo degradation time was 4 hours. The entire treatment period was not more than 4 days. From**Table 1**, it shows that the present method is more powerful than years that made biodegradation directly on beach [17,18].

DISCUSSION

The advantages of centralized treatment plant

The present scheme of centralized treatment plant has many advantages:

- Elimination of adverse environmental factors from the remediation process, such as the cold, waves, tides and human foibles. Because the oily pollutant has been transported to a centralized treatment plant where all conditions, such as temperature, pressure, humidity, fertilizer, water or air supply can be controlled, degradation becomes quite simple.
- Dramatic cost reduction, the cost of employees, transportation, material and utilities can be reduced because the working site is confined within the plant. Workers need not haul equipment and material to the beach, exempting it from corrosion, destruction and burglary.
- Managerial control, in a plant environment, workers efforts are focused upon the tasks at hand, thus saving them from wasted trips to and from beach. Treatment efficiency is greatly enhanced because progress and information are all controlled in the plant, minimizing the likelihood of mistakes.
- Rapid beach recovery, currently, the site for applications of the traditional method of degradation is the polluted site itself. The beach is closed and the oil stays there for years until (unacceptable) degradation is finished. However, where a beach is polluted by oil, the thickness of the oil layer is general not greater than 20 or 30 cm. Excavators could easily scrape the oil layer off beach. In this manner, the beach landscape is immediately restored and returned to local tourism and the marine fish culture industry can resume their actions as soon as possible. Therefore, the treatment of oil pollution through utilization of a centralized treatment plant is far better for trade than the traditional on-site degradation method.
- Sharply reduced time for achieving degradation to safe levels. It's important to emphasize that the period of degradation in a plant is considerably shorter than the period of degradation at a beach. The period of degradation at a beach takes years because of the impediments caused by weather, temperature or the human foibles that caused or extend the deterioration. Workers are often repulsed by the work of having to clear away the rubbish or animal bodies left by the retreating tide, the excreta of birds and dogs, and by having to keep the environment clean during the degradation period. Where degradation is done in a centralized plant, however, operating efficiency will be high because all these factors are removed and degradation is insulated from external disturbances. All the factors such as temperature, pressure, fertilizer, water or air supply can be controlled in the plant.
- Elimination of secondary pollution, almost certainly, there is no secondary pollution because the oil is degraded to safe levels and is not buried or rushed into the ocean to cause a pollution problem elsewhere.
- This presented method also can be applied for large area pollution, irrespective of whether the organization that does the remediation and pollutant degradation is private or publicly owned, the polluted beach would be restored to original appearance in a few days if the centralized treatment plant system was adopted to treat beach oil pollution.

The quick treatment for rocky coast oil pollution

As for coastal rock pollution, the common treatment was to recover the oil floating on the water. But helplessly, governments always took a passive attitude toward confronting the dried oil stains, or oil that lay in the crevices of the rocks. Some strong governments sent workers on mission rubbing the oily rocks, but at a snail's pace. Moreover, they washed out the oily debris into the sea, which caused the oil to return to the coast as a new source of pollution. It was all nonsense.

On the face of it, the characteristics of coastal rock pollution and beach pollution are quite different. Governments suggested spraying dispersant onto the polluted rocks to clean them up. It seemed that the oil stains on the rocks were disappearing but part of the oil-water mixture simply drifted to sea as fish food to enter the food chain. Part of the mixture sunk down into the sea as a pollution source. Nonetheless, in practice the treatment of beach pollution can be extended to coastal rock pollution. In other words, by utilization of a centralized treatment plant we can treat coastal rock pollution the same way as we treat beach pollution. The treatment is as follows.

First, people recover the floating oil on the water where coastal rock pollution has occurred. Second, clean the heavy oil from the rocks by oil absorbent sheets. Third, brush away the dry oil stain standing on the underside of a rock by electric brush. Finally, gather the oily dust and transport it to centralized treatment plant to be degraded just like the oily sand mentioned above. In this manner, the reinstatement work of a rock coast could be completed much faster.

In short, the treatment of beach and rocky coast oil pollution can be united by the method of a centralized treatment plant.

Choice of Schemes

There are two available schemes to be chosen by governments for treating coastal oil pollution:

- Governments could establish cooperation with environmental remediation companies. The advantage of such a policy is that it is easy to manage by governments. The disadvantage is the difficulty of close supervision. Environmental remediation companies only show governmental supervisors photos as evidence of environmental recovery, but pollutants are still always buried to decrease costs, causing secondary pollution. Governments must face the challenges with respect to adequate supervision.
- A large country could establish an environmental organization and pollutant cleaning plants. The organization, consisting of professional and full time members, is responsible for treating all serious polluting accidents in the country. For a small contrary, governments could establish pollutant cleaning plants as well as a pollution treatment group. The members of the group, forming a part time committee in the nature of a task force, should have regular training in pollution treatment to raise their skills, and be called up for emergency remediation whenever oil pollution occurs. They should be immediately dispatched to polluted sites and convey pollutants to a plant to affect biodegradation. The benefits of this management method are low cost, less manpower required, higher efficiency, shorter recovery and minimal secondary pollution. The drawbacks are the complexity of management and difficulty in maintaining the plant.

CONCLUSION

Coastal oil pollution is a major issue in marine ecology and economy. The most important aspect of the environmental remediation is the workflow design. The importance of the design presented in this paper is that oily pollutants from an open site can be transported to a centralized treatment plant where all factors are easy to control. The present scheme of a centralized treatment plant has many advantages:

- Elimination of adverse conditions, degradation is made quite simple and stable.
- Dramatic cost reduction with respect to employees, transportation, material and utilities.
- Managerial control, the workers focus their attention on the tasks at hand within the plant, minimizing the likelihood of mistakes.
- Rapid beach recovery.

Sharply reduced time for achieving degradation to safe levels.

Elimination of secondary pollution.

The procedure for treating beach oil pollution is described as follows. First, wash out the pollutant at a centralized treatment plant. Second, return beach sand to the beach. Third, filter the oily waste water by multilayer oil absorbent sheets. Forth, incinerate the sheets. Finally, degrade the residual oil in the waste water.

The procedure for treating coastal rock oil pollution is similar to treating beach sand pollution.

The ideas for governments to establish cooperation with environmental remediation companies or establish an environmental remediation organization have both advantages and disadvantages. If governments would actually put these supervisory and managerial recommendations into practice, the coast will be kept clean and beautiful.

REFERENCES

1. Islam ShahidulMD, Masaru Tanaka (2004) Impacts of pollution on coastal and marine ecosystems including coastal and marine fisheries and approach for management: a review and synthesis. Marine pollution bulletin 48: 624-649.
2. Etkin Dagmar Schmidt(2000) Worldwide analysis of marine oil spill cleanup cost factors. Arctic and marine oilspill program technical seminar: 1.
3. Mathieu Ducros, Sylvie Wolf, Bernard Carpentier, and Marie-Christine Cacas(2013) Full Integration of Biodegradation Processes in Petroleum System Modeling.
4. Lin MC (2011) The advanced development of microbial degradation to oil pollution. The conference of salvage at sea: 97-106.
5. Ronald AtlasM (1981) Microbial degradation of

petroleum hydrocarbons: an environment perspective. Microbiological reviews 45: 180-209.

6. Leahy JG, Colwell RR (1990) Microbial degradation of hydrocarbons in the environment. Microbiol Rev 54: 305-315.
7. Zahed MA, Aziz HA, Isa MH, Mohajeri L (2010) Effect of initial oil concentration and dispersant on crude oil biodegradation in contaminated seawater. Bull Environ ContamToxicol 84: 438-442.
8. Gallego JL, Loredo J, Llamas JF, Vázquez F, Sánchez J (2001) Bioremediation of diesel-contaminated soils: evaluation of potential in situ techniques by study of bacterial degradation. Biodegradation 12: 325-335.
9. Wu TY, Mohammad AW, Jahim JM, Anuar N (2010) Pollution control technologies for the treatment of palm oil mill effluent (POME) through end-of-pipe processes. J Environ Manage 91: 1467-1490.
10. Lin MC, Kao JC (2016) Marine environmental protection: An application of the nanometer photo catalyst method on decomposition of benzene. Mar Pollut Bull.
11. Lin MC, Kao JC (2016) Marine environmental protection: An application of the nanometer photo catalyst method on decomposition of benzene. Mar Pollut Bull.
12. Lin MC, Chen JY, Jhuang JJ (2012) The advanced nanotechnology on treating marine light oil pollution. The 19th symposium of maritime police: 205-214.
13. Lin MC, Chen JY (2012) The application of nanotechnology to treatment of coastal oil pollution. The 2rd marine affairs conference-marine policy and management: 94-107.
14. Lin MC (2010) The physical model of marine oil pollution and its solution.The Underwater Technology Conference: 20-26.
15. Lin MC, Chuang KL (2013) The treatment and survey of marine oil pollution. The marine affairs and management conference: 161-176.
16. Wilfred FMR,Michael GM,Martin DJ,Fratepietro F,Swannell RPJ (2004) Bacterial Community Dynamics and Hydrocarbon Degradation during a Field-Scale Evaluation of Bioremediation on a Mudflat Beach Contaminated with Buried Oil.Appl Enviro Microbiol70): 2603-2613.
17. Jiménez N, Viñas M, Bayona JM, Albaiges J, Solanas AM (2007) The Prestige oil spill: bacterial community dynamics during a field biostimulation assay. Applied Microbiology and Biotechnology 77: 935-945.
18. ChaerunSK, Tazaki K, Asada R, Kogure K (2004) Bioremediation of coastal areas 5 years after the Nakhodka oil spill in the Sea of Japan: isolation and characterization of hydrocarbon-degrading bacteria. Environment International 30: 911-922.

CHAPTER 10

REMOTE SENSING OF THE COASTAL ECOSYSTEMS

Palanisamy Shanmugam

Department of Ocean Engineering, Indian Institute of Technology Madras, Chennai 600036, India

INTRODUCTION

The coastal ecosystems comprised of biological, physical and chemical components, are indispensable to life on our planet and support the livelihood of people living and depending on coastal resources. These ecosystems are one of the most threatened but mostly populated landscapes in the world, which can no longer provide their biological functions and regulate services that sustain coastal economic production and livelihoods [1]. Thus, there is a pressing need for plans to address issues such as protection and sustainable use of coastal ecosystems. It is also a major concern of international organizations, including the Food and Agricultural organization (FAO), International Union for Conservation of Nature (IUCN), the UN Commission on Sustainable Development (CSD), Ramsar Convention of Wetlands and the Convention on Biological Diversity. It thus becomes very essential for maintenance and management of ecosystems, such as wetlands, mangrove swamps, salt marshes, estuaries, bay, and coral reefs that are home to many different types of plants and animals and support human needs. The ecosystem approach is a strategy for the integrated management of land, water and living resources that provides sustainable delivery of ecosystem services in an equitable way, thus its management is to be designed and executed in an adaptive, learningbased process that applies the principles of the scientific method to the process of management [2].

Coastal zones are the site of complex natural systems where intense interactions occur between the hydrosphere, lithosphere, and the atmosphere, which meet and interact, forming interconnected systems [3]. It is to be noted that healthy coastal ecosystems provide valuable services such as food security, resources for economic growth, recreation alongside tourism, coastline protection and crucial reservoirs of biodiversity [2]. Global climatic and oceanographic events contribute to sea level rise, severe storm events, coastal erosion, and sea surface temperature, all of which can harm coastal ecosystems. In addition to the direct impacts coastal areas such as barrier islands, beaches, and wetlands are highly sensitive to sea-level changes. Rising seas will intensify coastal flooding and increase the erosion of beaches, and wetlands, as well as threaten jetties, piers, seawalls, harbors, and waterfront property [4]. Increasing stress keeps accelerating human interventions that impact natural coastal ecosystem processes and it becomes paramount importance for environmental monitoring and assessment in order to manage and protect these sensitive areas more effectively through continues and coordinated observations. Understanding of regional and global-scale variability of coastal ecosystem has been limited by in-situ sampling confined to point measurements. Remote

Citation: Shanmugam P (2013) Remote Sensing of the Coastal Ecosystems. J Geophys Remote Sensing S2:e001. doi:10.4172/2169-0049.S2-e001

sensing technique is the best identified alternatives and the most cost-effective method for providing temporally resolved synoptic views of coastal zone region over long periods of time.

Coastal ecosystems reveal tremendous variations in aerial extent, spatial complexity, and temporal variability, thus required to be observed with spatial, spectral, and temporal resolutions from both satellites and aircrafts. Some of the ecosystem health indicators that can be observed by remote sensors include wetland loss and fragmentation, wetland biomass change, vegetation cover, deforestation of mangroves, percentage of impervious watershed area, buffer degradation, and changes in hydrology, changes in shoreline and high water line, erosion and accretion, harmful algal blooms, water turbidity, chlorophyll concentration, eutrophication level, salinity and temperature [4-12]. Protecting coastal ecosystems requires the ability to monitor changes in their biophysical features and controlling processes at high spatial and temporal resolutions. At present satellite and airborne remote sensors can map and measure coastal ecosystems and their changes cost effectively at suitable scales and resolutions, minimizing the need for extensive field and ship measurements [4]. This special issue demonstrates the use of cost-effective remote sensing technologies and related field observation techniques for evaluating, monitoring and sustainable management of coastal ecosystems. This issue evaluates the major types of global coastal ecosystems, discusses the major benefits and possible challenges of using remote sensing geophysical technologies. It covers a variety of remote sensing techniques, such as multispectral, hyperspectral imaging spectrometers, thermal infrared sensors, synthetic aperture radars, scatterometers, altimeters and LIDAR. It emphasizes not only the integration of data acquired from these sensors on various types space-borne and air-borne platforms to provide sufficient spatial, spectral, radiometric and temporal resolutions, but also the incorporation of a reliable field data collection approach to acquire sufficient in-situ measurements that are used to calibrate and validate the remotely sensed information.

Content of the Special Issue

This special issue is targeted for scientists, managers and researchers interested in using remote sensing in the study or management of coastal ecosystem and would serve as an important reference source for the scientific community.

This special issue consolidates various papers that address diverse topics of coastal ecosystem [1-12], such as:

- Monitoring coastal wetlands and watersheds by studying the impact of land runoff on estuarine and coastal ecosystems, including watershed models, hydrodynamic models, water quality models, and living resource models
- Examine wetland changes and land cover trends, such as impact of rising sea levels and hurricanes on wetlands, by means of analyzing time series of remotely sensed imagery.
- Understanding shoreline topography and bathymetry, since topographic and hydrographic information are basic elements in studies of near shore geomorphology, hydrology, and sedimentary processes.
- Mapping submerged aquatic vegetation, coral reefs, and general bottom characteristics becausedecline of coral reefs is closely linked to human activity, as they thrive in a narrow range of environmental conditions and are very sensitive to small changes in temperature, light, water quality, and hydrodynamics.
- Monitoring of ocean chlorophyll since biological productivity can be estimated by measuring the chlorophyll-a concentration.
- Observing eutrophication and hazardous algal blooms in view of the fact that high concentrations of nutrients exported from agriculture or urban cover in coastal watersheds, or produced by coastal upwelling, are causing algal blooms in many estuaries and coastal waters.
- Supervising sea surface temperature (SST) and sea surface salinity (SSS) because they are necessary for estimating the source of heat at the air–sea boundary. Sea surface temperature and salinity are also important input data for estimating the steric component of the sea level, which is important for coastal studies. SSS is critical for determining the global water balance, for understanding ocean currents, and for estimating evaporation rates.
- Detection and tracking of oil spill, so as to limit the damage by a spill and facilitate containment and cleanup efforts. Also the environmental engineers and managers and other responsible scientific community obtain information on spill location, size and extent of the spill, direction and speed of oil movement, and wind, current, and wave information for predicting future oil drift and dispersion.
- Studying physical offshore processes as the currents and waves that strongly affect coastal ecosystems, which is an extremely dynamic environment.

PERSPECTIVES

By virtual of their position, coastal areas are highly dynamic and located at the interface between terrestrial and aquatic ecosystems, which are one of the most challenging targets for environmental remote sensing. This issue presents the latest developments in the use of remote sensing and geospatial technologies for characterizing coastal wetlands and watersheds, shorelines, coastal waters, submerged aquatic vegetation, benthic habitats, and coral reefs. Case studies of the operational use of remote sensing in evaluating changes, ecosystem monitoring and management are dealt with. The different applications in this issue are representative of the interdisciplinary, multi-scale studies taking advantage of remote-sensing capacities for coastal science, management and conservation. Recently launched satellites are providing the means for more accurately detecting changes in coastal ecosystem health, including biological productivity and habitat quality. Thermal IR imagers and new satellite microwave radiometers are able to map sea surface temperatures, salinity, and soil moisture. Scatterometers, altimeters, and SAR data are providing more accurate information on sea surface winds, elevation, currents, wave fields and oil slicks and oil spill drift and dispersion which can be used in the models to predict their impacts on coastal ecosystems. Even such voluminous data is insufficient, in dealing with problems related to the coastal ecosystem. However, the above perspectives give tremendous scope for practical experimentation to take up research related to remote sensing of coastal ecosystems. Paper presented in this special issue is a reflection of tedious and worthwhile data collected at various time steps.

REFERENCES

1. Wattage P (2011) Valuation of ecosystem services in Coastal Ecosystems: Asian and European Perspectives. The United Nations Environment Programme, Nairobi, Kenya.
2. UNEP (2011) Taking Steps toward Marine and Coastal Ecosystem-Based Management-An Introductory Guide.
3. Wilches G (1995) Disasters and the environment. (2nd Edn).
4. Klemas V (2011) Remote sensing techniques for studying coastal ecosystems: an overview. Journal of Coastal Research 27: 2-17.
5. Lathrop RG, Cole MB, Showalter RD (2000) Quantifying the habitat structure and spatial pattern of New Jersey (U.S.A.) salt marshes under different management regimes. Wetlands Ecology and Management 8: 163-172.
6. Richardson LL, LeDrew EF (2006) Remote Sensing of Aquatic Coastal Ecosystem Processes: Science and Management Applications. Remote Sensing and Digital Image Processing 9: 330.
7. Shanmugam P, Ahn YH, Sanjeevi S (2006) A comparison of the classification of wetland characteristics by linear spectral mixture modeling and traditional hard classifiers on multispectral remotely sensed imagery in southern India. Ecological Modelling 194: 379-394.
8. Shanmugam P, Ahn YH, Sanjeevi S, Manjunath AS (2003) Integration of ERS- 2 SAR and IRS- 1D LISS-III for improved coastal wetland mapping. Korean Journal of Remote Sensing 19: 351-361.
9. Shanmugam P, Ahn YH, Ram PS (2008) SeaWiFS sensing of hazardous algal blooms and their underlying physical mechanisms in shelf-slope waters of the Northwest Pacific during summer. Remote Sensing of Environment 112: 3248- 3270.
10. Shanmugam P (2011) A new bio-optical algorithm for the remote sensing. Journal of Geophysical Research 116: 12.
11. Ahn YH, Shanmugam P, Moon JE, Ryu JH (2008) Satellite remote sensing of a low-salinity water plume in the East China Sea. Annales Geophysicae 26: 2019-2035.
12. Ahn YH, Shanmugam P, Gallegos S (2004) Evolution of suspended sediment patterns in the East China and Yellow Seas. Journal of the Korean Society of Oceanography 39: 26-34.

CHAPTER 11

THE ROLE OF REMOTE SENSING AND GIS FOR LOCATING SUITABLE MANGROVE PLANTATION SITES ALONG THE SOUTHERN SAUDI ARABIAN RED SEA COAST

Ibrahim H. Elsebaie, Ayman S. H. Aguib, Dafer Al Garni

Civil Engineering Department, College of Engineering, King Saud University, Riyadh, KSA

ABSTRACT

This paper investigated a methodology for locating suitable sites for establishing mangrove plantations along the south Saudi Arabian Red Sea coast based on Geological characteristics, geomorphology and discharge of drainage effluents as primary sitting criteria. Remote sensing Satellite imagery and digital elevation models were interpreted to determine most of the parameters, the rest were measured in field. In the study, it was found that some healthy mangrove stands now exist along the Red Sea coast near the city of Jizan although many stands have been destroyed in recent years due to rapid development, infilling and over-grazing and use of the mangroves as fuel. Remote sensing satellite imagery was used to map exiting Mangrove stands of different densities along the study area. The study determined that mangrove stands are primarily concentrated in the southern part of the study area and that they are sparsely found northward. Interpretation of A Landsat Thematic Mapper panchromatic band image and DEM showing that red sea coast along the study area are mainly covered by fans of ten drainage basins where the Flood potentiality of these basins was determined. The study was able to determine that three areas, which have a moderate to low flash flood potential, have the necessary environmental elements necessary to support mangroves. One, in particular, would be the best site to establish a mangrove plantation project.

INTRODUCTION

Mangroves play a vital role in supporting the surrounding environment. Mangroves play an important role in controlling erosion and protecting the coastline. Along the Red Sea coast, mangrove stands are distributed. Since the mid-1980s, these stands have been threatened by rapid tourist development and grazing by Bedouins. The size and the number of these stands have been reduced and many of the stands have been destroyed completely [1,2]. Thus, it is necessary to conserve the existing mangroves stands and plant mangroves where ever they can be grown. Mangroves protect shorelines from erosion by stabilizing sediments with their tangled root systems. They maintain water quality and clarity, filtering pollutants and trapping sediments originating from land. Mangroves function as nurseries for shrimp and recreational fisheries, exporters of organic matter to adjacent coastal food chains, and enormous sources of valuable nutrients [3].

Citation: I. Elsebaie, A. Aguib and D. Garni, "The Role of Remote Sensing and GIS for Locating Suitable Mangrove Plantation Sites along the Southern Saudi Arabian Red Sea Coast," International Journal of Geosciences, Vol. 4 No. 2, 2013, pp. 471-479. doi: 10.4236/ijg.2013.42044.

Mangroves occur in sparse, but natural, formation along the Saudi Arabian Red Sea coast. Many more mangroves stands existed, but have been destroyed by factors such as overgrazing by Bedouins who use the trees as a fuel and, more recently, by development. Saifullah [4] presented an overview of various mangrove studies along the Red Sea coast of Saudi Arabia and concluded that information on the subject is inadequate.

There are many recent studies on the ecology of mangroves and setting up mangrove plantations [5,6]. The consensus of these authors is that mangroves are threatened around the world due to infilling and other forms of development and pollution.

There is a large body of scientific literature on using remote sensing for assessing the condition of mangroves [7-12]. Many other studies have used remote sensing for mapping mangroves and locating suitable sites for mangrove plantations as reported in [13-16].

The distribution of mangroves in both northern and southern Red Sea coast of Saudi Arabia is patchy and they do not occur as a continuous forest and are generally restricted to the low energy quieter environments. Mangroves occupy similar environments in both regions and are under similar stresses caused by various anthropogenic activities and camel grazing. However, mangrove growth in south is relatively denser in comparison to north due mainly to the tropical climate, more wadis, higher rainfall and nutrients and less saline waters [2].

In this study, sites suitable for mangrove plantations along the southern Saudi Arabian Red Sea coast were identified. The study demonstrated the use of remote sensing and geographic information systems to identify suitable locations to establish mangrove plantations. The present study focused on the critical environmental factors necessary for successful mangrove plantation development.

GEOGRAPHICAL LOCATION OF THE STUDY AREA

The study area lies between 16°01'N and 19°57'N and between 40°38'E and 43°55'E as shown in **Figure 1**. Jizan city lies in the western southern part of the kingdom of Saudi Arabia on the Red Sea Coast; this city is characterized by rainfall storms which are varying in intensity and duration. The southern part of the city is sometimes exposed to the risk of flash flooding due to the heavy rainfall intensity which may happen during the rainy storms, the peak runoff flows from the East towards West.

The highest concentrations of mangroves are found near the coastal city of Jizan. North and south of Jizan are long stretches of coastline barren of any type of vegetation and stretches with relatively small patches of mangroves.

Jizan has great tourism potential owing to the abundance of near-shore coral reefs, natural mangroves and beautiful oasis. The Farasan Islands which as 40 km from Jizan, are a protected series of coral islands and are habitat to endangered species of many animals. Jizan has grown from a village of fewer than 10,000 people in

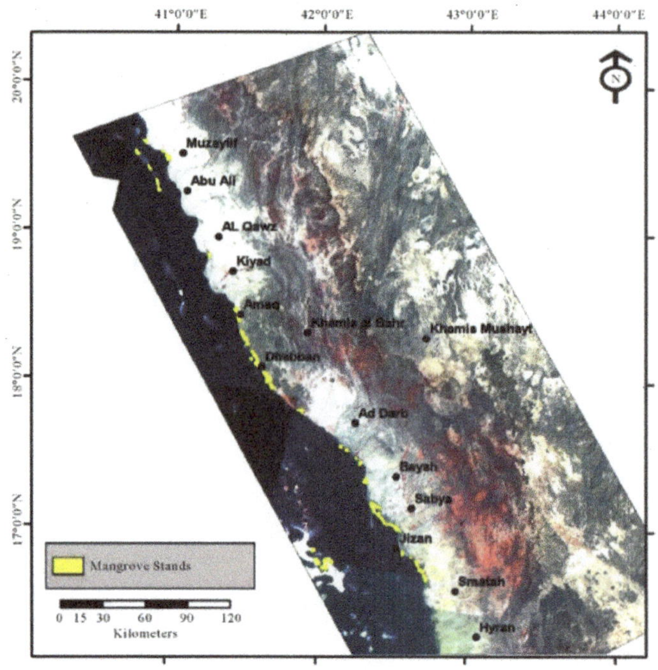

Figure 1. Mangrove stands as interpreted from GeoEye images showing location of the study area.

1978 to over 100,000 by 2008 and so care must be taken to protect the mangroves from development.

MATERIALS AND METHODS

The six critical environmental factors that determine if a coastline is capable of starting and maintaining a mangrove plantation are: 1) air temperature; 2) water temperature; 3) water salinity; 4) tidal and wave energy; and 5) flood risk; and 6) soil type and stability [17]. The suitability value ranging for these environmental parameters are:

Air temperature: 8 - 47 degrees C; Water temperature: 22 - 34 degrees C; Water salinity: 33.8%; Wave and tide energy: low; Flood potential: low; and Soil type: Sandy to sandy loam. With these environmental boundaries in mind, the objective was to determine where, if anywhere, on the southern Saudi Arabian Red Sea coast these environmental requirements were met.

Existing Mangrove Stands and Primary Selection of the Sites

The existing Mangrove Stands along as interpreted from GeoEye satellite images of the period 2005 to 2008 by visual screen digitizing. The identified stands checked and corrected at the field. The Mangrove stands are shown in (**Figure 1**).

Mangrove can only grow in areas away from direct waves. Satellite images of the area extended along the Saudi Red Sea shoreline, starting from Saudi Arabia— Yemen boundary up to 300 km north had been studied to define the coastal area protected from direct current waves. These areas include small bays or area protected from the waves by islands.

Ten locations meet the searching criterion were sited. All these locations already have mangrove stand(s) (See **Figure 2**). So that, in this study our objective will be, selecting the best suitable site for mangrove plantation

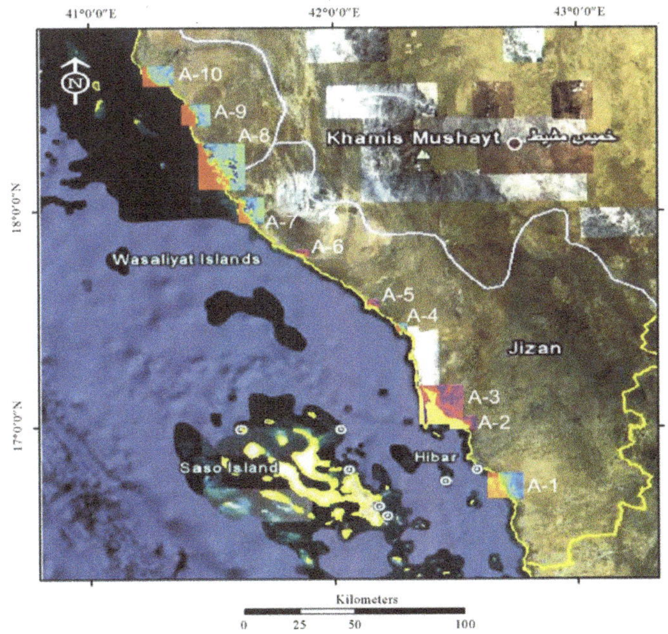

Figure 2. Location map of the ten sites protected from direct current waves in the study area.

out of these ten locations instead of finding a new location. It is impossible to have Mangrove plantation project in open sea area. The best location for mangrove plantation project is the location with low mangrove density (not related to environmental aspects), good soil and away from urban area and basins with high flood potentiality.

Water Temperature

Data from the U.S. National Oceanographic and Atmo= spheric Administration (NOAA) on the average yearround near-shore sea temperature and salinity for the near-shore zone in the study area were obtained. Water temperature (surface) were derived from NOAA's Advanced Very High Resolution Radiometer (AVHRR) satellite which is accurate to within 0.5 degrees C. Avicenna marina, the dominant mangroves species naturally found in the Red Sea coast requires relatively warm water [18].

The average sea surface temperature of the study area ranges between 26°C in winter and 32°C in summer is suitable for mangroves. The temperature varies by approximately 2°C between the north and south boundary of the study area. So, temperature is not an important factor in site selection in this study.

Air Temperature

Mangroves tolerate air temperatures as low as 8 degrees C for short periods of time and as high as 42 degrees C [18]. They thrive in the 28 degree to 30 degree temperature range. Data was obtained from the Presidency of Meteorology and Environment, Kingdom of Saudi Arabia for meteorological data including monthly air temperature for the months of January and August, representing the coldest and the hottest months, respectively, were obtained for the years 1999-2010. The lowest recorded temperature for the study area was 20 degrees C and the highest was 36 degrees C. These temperature extremes are well within the range tolerated by mangroves.

Salinity

The Red Sea is one of the most saline bodies of water in the world, due to high evaporation. Salinity ranges from between ~36 ppt in the southern part due to the effect of the Gulf of Aden water and reaches 41 ppt in the northern part, due mainly to the Gulf of Suez water and the high evaporation. The average salinity is 40.

Mangrove trees are relatively tolerant of salt. They are able to withstand concentrations of up to 39 ppt (part per thousands) and thrive in concentrations of about 36 ppt [19].

The average salinity concentration for the study area, according to NASA [20] is between 36% to 38%. Therefore, salinity is not a limiting factor for mangroves.

The southern part of the study area has lower salinity (36% to 37%) than the northern part (37% to 38.5%). Although salinity is suitable for Mangrove all over the study area, but Mangroves can grow better in the southern part that has a lower salinity, this might explain the high dense Mangrove stands existing in Gizan area.

Remote-Sensing and GIS Applications

Data for GIS were based on existing available sources in the study area that include satellite images acquired on different dates. Digitized map of the study area has been prepared. Multiple resource uses data were collected during the field trips and were used as bases in the selection of training areas. Old records, management history and other information (during the field trips) were collected from government offices and local people. Geometric correction was performed with fourth order polynomial and the root mean square (RMS) for the ground control points (GCP). One of the Landsat ETM images which were used in the study is shown in (**Figure 3**).

The software ArcGIS was used to define the drainage network of the study area using USGS 7.5 minute (30- by 30-m data spacing) Digital Elevation Models (DEM's). The USGS 7.5-minute DEM cast on Universal Transverse Mercator (UTM) projection. A Landsat Thematic Mapper panchromatic band image was used to check and correct the drainage network produced by ArcGIS.

The River Tools software program was used to generate a grid of the drainage patterns of all of the basins in the study area. Using ArcGIS, the grid was converted into a shapefile. The shapefile was then used to determine the morphological features (i.e. order of each individual segment of the basin and its length, basin length, basin shape, drainage density, stream frequency and bifurcation ratio) of each watershed. These factors, in addition to basin slope were determined using a digital elevation model (DEM) and the three dimension models of the study area as shown in Figures 4 and 5. These models were critical in determining the basin's proneness to flash flooding.

Basin length is the length measured along the principal flow path from the watershed outlet to the basin boundary. It is a measure of the travel time of water through the basin. Basin shape indicates the path runoff will travel through various parts of the basin and where runoff will discharge from the outlet.

Drainage density and stream frequency are the total length and total number of the entire stream segments in a drainage basin divided by the total area of the drainage basin. High drainage density or/and stream frequency of a basin indicates a rapid storm response and conesquently high probability to result in a flash flood. The bifurcation ratio is the ratio of the number of streams of any order to the number of streams of the next highest order. A lower bifurcation ratio indicates a circular basin with high flood potentiality and visa a versa. In order to determine the bifurcation ratio, the steam order was determined for the entire segment in each basin using the Horton Law [21].

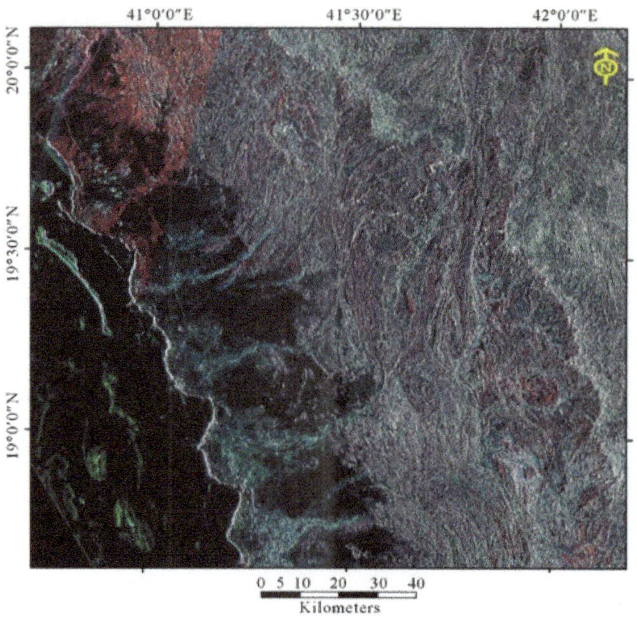

Figure 3. LANDSAT ETM images (bands 4, 3 and 2).

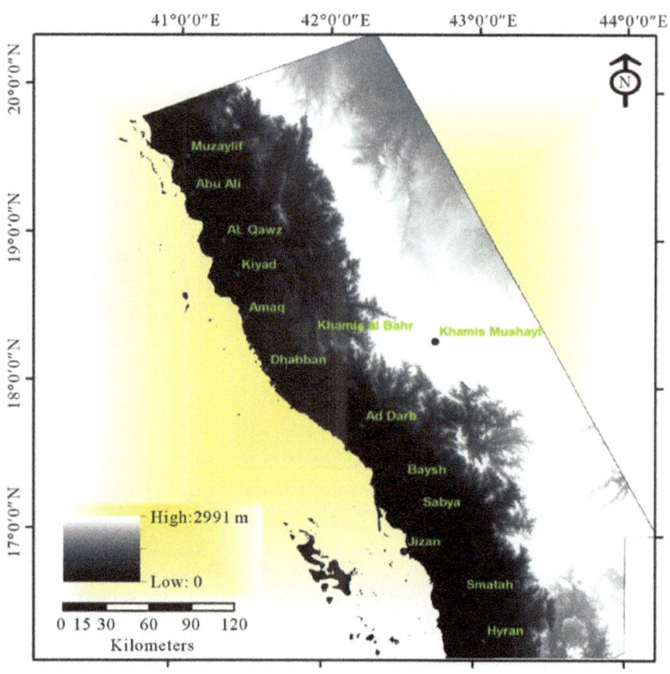

Figure 4. Digital elevation model of the study area.

Figure 5. The three-dimension models of the study area.

Basin slope is the average slope of a basin and is the most important variable affecting the momentum, speed and concentration of runoff. The possibility of starting a runoff event increases with increasing slope.

Some areas of the Saudi Arabian Red Sea coast are high wave energy zones due to their position with respect to the prevailing winds and also a lack of barrier islands which dissipate wave energy. They would be unsuitable locations for mangrove plantations. Other areas have barrier islands, protective bays and/or wide flat beaches which would be suitable from the wave perspective.

In order to identify zones along the study area that might have high wave energy, a three-dimension model of the study area was constructed. The model allowed us to identify zones with deltas with tidal creeks, small bays and channels with wide flat shores. These topographic conditions are correlated with low wave energy. The study area is classified into three classes with respect to wave energy: high, moderate or low as shown in (**Figure 6**).

RESULTS AND DISCUSSION

Spot and GeoEye remote sensing satellite images that available for the study area were obtained. SPOT and GeoEye image were interpreted to map the existing mangrove stands. On GeoEye images, discrimination between different densities of mangroves was observed and mapped. Using the SPOT imagery, the presence or absence of mangroves was mapped.

The factors considered for suitability of a site for mangrove plantation were sea water surface temperature, seawater salinity, tidal and wave energy, flood risk potentiality, soil type, soil pH, soil salinity and soil texture. Salinity and surface water temperature were found suitable for Mangrove along the study area.

Flash floods pose a danger to young mangrove trees because they can cause the seedlings to uproot and carry them into the sea. Floods occur when precipitation falls too quickly on saturated soil or dry soil that has poor absorption ability due to its clay content. The runoff collects in low-lying areas and rapidly flows downhill. Flash floods occur in normally dry areas that have recently received precipitation, but may be seen anywhere downstream from the source of the precipitation, even many kilometers from the source. Flash floods are common in the arid plains of Saudi Arabia where there are large drainage basins that feed into one tertiary channel.

The best sites to actively seed mangrove plants are those that are sheltered from wave energy and have gently sloping foreshores. Optimum growth and establishment success appears to occur at one meter from the highest tidal zone down to the lowest tidal line. In Kuwait mangrove plantation project, the Bahrain ecotype of Avicennia marina—the species recommended for our area of study—the majority of the seedlings planted in the sandy area above the tidal line survived and produced spreading plants with luxuriant foliage and dense canopies. In contrast, only a handful of the seedlings survived below the tidal line and grew poorly [17].

Based on interpretation of the DEM and the calculations of basin length and shape, drainage density, stream frequency and bifurcation ratio, three basins with low wave energy and low flash flood potential were selected for being suitable for mangrove plantations. They were the Al-Ahsabah, Yabah and Kid basins and are shown in (**Figure 6**).

A number of basic soil parameters can be determined from these grain-size distribution.

Once three sites (Al-Ahsabah, Yabah and Kid) were selected based their and water temperature, salinity, low wave energy and low chance for flash flooding, soil samples were collected from each site. Coordinates for each sample location were determined using a surveygrade GPS. Four samples were taken at each site and divided into depth increments of 0 - 20, 20z - 40, and 40 - 60 cm.

Mangroves grow in soil with high sand content, low silt and clay fractions and low calcium carbonate content [17]. Soil characteristics for the study area are a function primarily of the type of local exposed rock and soil transported from the basins that drain to the Red Sea. In order to assess the capability of each basin with respect to soil, the following variables were assessed: Lineament

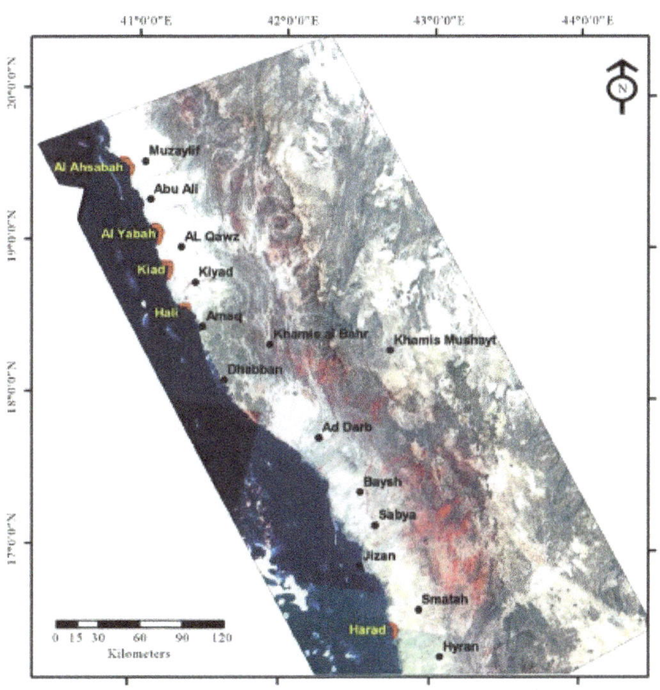

Figure 6. Areas with low wave and tide energy.

intensity; Drainage density; Area of exposed rocks and their mineral composition; Basin order; Basin bifurcation ratio; Drainage density; and Slope.

The formations and rock groups exposed on the surface were ranked based on the mineral composition of their rocks as: 1) very good; 2) good; 3) moderate; and 4) poor. The soil suitability factor (SSF) was determined for each exposed rock unit within the basin based on the variables assessed before.

The soil suitability factor was determined for each basin by averaging those of all rock unit exposed within the basin and then re-scaling the entire area in a range of between 1 and 100. The SSF was highest for the Yabah site. Therefore, based on all the criteria we measured and assessed, the Yabah site is the most suitable for establishing a mangrove plantation in the study area. Plantations could also be established at the other two sites, Al-Ahsabah and Kid, because they possess all the environmental characteristics required for mangroves.

CONCLUSIONS

Dense mangrove stands are found along the Red Sea coast in Jizan, some of these stands have been destroyed in recent years. This loss was the result due to development, infilling, tourist activities, over-grazing and use of the mangroves as fuel. There was limited evidence that the places where low-density mangrove stands now exist could not support high density stands of mangroves. Indeed, there was evidence that dense, healthy mangroves existed in these areas in the past and so could in the future.

Environmental factors, including temperature, salinity and rainfall also important key that have a large influence over the growth, survival and distribution on mangroves. Also Soil type is critical for successful or mangroves.

Selecting the sites suitable for mangrove plantation based on sea water surface temperature, seawater salinity, tidal and wave energy, flood risk potentiality, soil type, soil pH, soil salinity and soil texture.

It can be seen from (**Figure 7**) that two basins have high flood potentiality Hali and Tashar. Although the rest of basins specially those of large areal extent such as Wadi Yabah, Al Ahsabah, and Qanunah have moderate to low flash flood potential. Therefore, the Hali and Harad sites are risky for flash floods that can destroy the mangrove propagules in the early plantation stage. Therefore, it is recommended to place the mangrove plantation site away from the main outlet of the basins of high flood potential, especially when it is accompanied with high slope such as Wadi Hali and Tashar. For that reason Hali and Harad sites were excluded from further assessment.

There are three sites in the study area suitable for establishing mangrove plantations. They are shown in (**Figure 7**). The Yabah basin is the best site of the three. It has all the physical characteristics required for successful mangrove growth.

The study found that remote sensing imagery was very useful for mapping critical features such as drainage pattern and existing mangrove stands. While these features could be mapped on the ground, remote sensing was far more efficient and accurate.

ACKNOWLEDGEMENTS

Funding by the long term comprehensive National Plan for Science, Technology and Innovation, KACST, for the

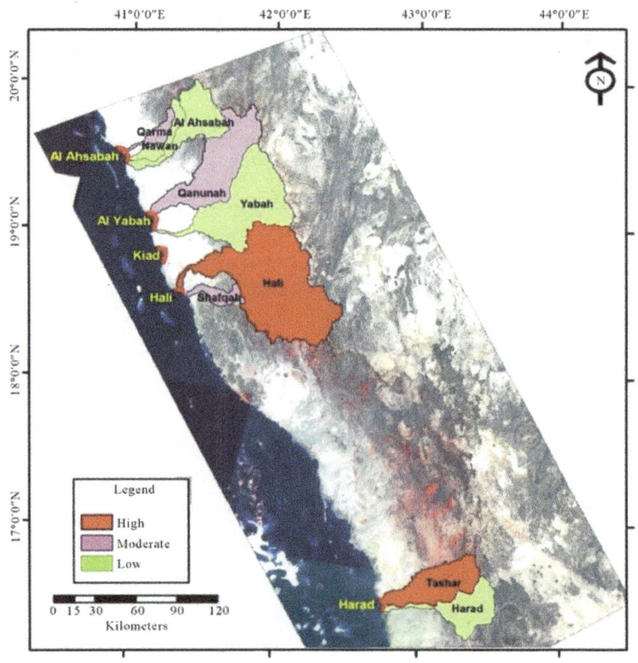

Figure 7. Flash flood potentiality.

research is gratefully acknowledged. We are also grateful to the Environmental Affairs office in Jizan, and Hurghadah for their helpful information, the valuable photos and studies.

REFERENCES

1. L. El-Juhany, "Present Status and Degradation Trends of Mangrove Forests on the Southern Red Sea Coast of Saudi Arabia," American-Eurasian Journal of Agricultural and Environmental Science, Vol. 6, No. 3, 2009, pp. 328-340.
2. M. Khan, A. Kumar and A. Muqtadir, "Distribution of Mangroves along the Red Sea Coast of the Arabian Peninsula: Part 2. The Southern Coast of Western Saudi Arabia," Journal Earth Science India, Vol. 3, 2010, pp. 154- 162.
3. H. El Monsef and S. Smith, "Site Selection for Mangrove Plantations along the Egyptian Red Sea Coast," World Applied Sciences Journal, Vol. 3, No. 7, 2008, pp. 740-747.
4. S. M. Saifullah, "Mangrove Ecosystem of Saudi Arabian Red Sea Coast—An Overview," Journal of King Abdulaziz University, Marine Science, Special Issue on Red Sea Marine Environment, Vol. 7, 1996, pp. 263-270.
5. R. Lewis, "Ecological Engineering for Successful Management and Restoration of Mangrove Forest," Ecological Engineering, Vol. 24, No. 4, 2004, pp. 403-418.
6. L. Gnanappazham, "Mangroves of Godavari—Anaylsis through Remote Sensing Approach," Wetland Ecology and Management Journal, Vol. 8, No. 2, 2006, pp. 89-94.
7. R. Krishnamoorthy, "Remote Sensing for the Assessment and Management of Mangroves: Indian Experience," Application of Remote Sensing and GIS for sustainable Development at National Remote Sensing Agency, Hyderabad, 1997.
8. T., Kannan, T. Ajith Kumar and A. Duraisamy, "Remote Sensing for Mangrove Forest Management," CAS in Marine Biology, Annamalai University, Tamil Nadu, 2008.
9. Y. Wang and A. Ngusaru, "Remote Sensing of Mangrove Change along the Tanzania Coast," Marine Geodesy, Vol. 26, No. 1-2, 2003, pp. 1-14. doi:10.1080/01490410306708
10. S. Ongsomwang, A. Rangsipanich, S. Punkul and J. Suekuni, "Evaluating Mangrove Plantation Sites Using Remote Sensing and GIS in Surat Thani Province," GIS/RS Based Assessment of Mangrove Resources—Final Report, Bangkok, 2007, 47 p.
11. M. Schwartz, "Mangroves: Remote Sensing," Encyclopedia of Coastal Science, Vol. 2, Springer Publishers, Berlin, 2005, pp. 614-621.
12. T. Truong, H. Pham, T. Thoab and L. Nguyen "Using Multi-Temproal Remote Sensing Data to Manage the Mangrove for Coastal Environmental Protection," The International Archives of the Photogrammetry, Remote Sensing and Spatial Information Sciences, Vol. 37, Beijing, 2008, pp. 709-711.
13. E. Green, C. Clark, P. Mumby and A. Ellis, "Remote Sensing Techniques for Mangrove Mapping," International Journal of Remote Sensing, Vol. 19, No. 5, 1998, pp. 935-956.doi:10.1080/014311698215801
14. T. Coleman, A. Mana and Y. Twumansi, "Application of Landsat to Study of Mangrove Ecology along the Coast of Ghana," Journal of the ISPRS, Vol. 49, 2008, pp. 362-379.
15. N. Azlan and R. Othman, "Monitoring of Mangrove Areas Using Remote Sensing toward Shoreline Protection," Proceedings of GIS Osrava, Malaysia, 2009.
16. F. Howari, D. Jordan, N. Bouhauche and S. WyllieEcheverra, "Field and Remote Sensing Assessment of Mangrove Forests and Seagrass Beds in Northwest United Arab Emirates," Journal of Coastal Research, Vol. 25, No. 1, 2009, pp. 48-56. doi:10.2112/07-0867.1
17. N. Bhat and M. Suleiman, "Classification of Soils Supporting Mangrove Plantation in Kuwait," Archives of Agronomy and Soil Science, Vol. 50, No. 6, 2004, pp. 535- 551.doi:10.1080/03650340410001729726
18. H. Teas, "Biology and Ecology of Mangroves," W. Junk Publishers, The Hague, 1984, 188 p.
19. N. Schmitz, V. Anouk, H. Beekman, J. Gitundu and N. Koedam, "Influence of a Salinity Gradient on the Vessel Characters of the Mangrove Species Rhizophora Mucronata," Annals of Botany, Vol. 98, No. 6, 2006, pp. 1321- 1330. doi:10.1093/aob/mcl224
20. NASA, 2012. http://science.nasa.gov/earth-science/oceanography/physical-ocean/salinity/
21. D. Eash, "A Geographic Information System Procedure to Quantify Drainage-Basin Characteristics," Journal of the American Water Resources Association, Vol. 30, No. 1, 2007, pp. 1-8.

CHAPTER 12

COASTAL VULNERABILITY AND SHORELINE CHANGES FOR SOUTHERN TIP OF INDIA-REMOTE SENSING AND GIS APPROACH

V. Joevivek[1,2], S. Saravanan[1] and N. Chandrasekar[1]

[1]Centre for GeoTechnology, Manonmaniam Sundaranar University, Tirunelveli 627102, India

[2]Civil engineering department, Christ the king engineering college, Karamadai, Coimbatore 641104, India

ABSTRACT

The present research aims to classify the vulnerable risk zones of the Southern tip of India using shoreline change analysis and coastal vulnerability index (CVI). The shoreline change analysis has been done by automatic image analysis techniques using multi-temporal Landsat data (1973, 1992, 2000 and 2006). The results have shown remarkable erosion was found in Cape Camoron (Left) (4.21 m/year), Idindakarai (4.56 m/year) and Vijayapathi (4.66 m/ year). In contrast, the station between Chinna muttam and Visvanarayanapuram has predominant deposition. The CVI index were established based on following six variables: Geomorphology, Shoreline change rate (m/yr), Coastal slope (deg), Relative Sea level change (mm/yr), Mean wave height (m), Mean Tide range (m). According to the CVI value, Cape Camoron (Left), Idindakarai and Vijayapathi sites are identified as highly vulnerable zones. Overall the study, remarkable coastal landform dynamics observed in chinna muttam. The vulnerable map prepared for the southern tip of Indian coast can be useful to prevent the coastline erosion and future disaster mitigation.

INTRODUCTION

The dynamics of coastal landforms is mainly controlled by nearshore processes, beach morphology and anthropogenic activities. The unusual natural disasters and continuous engineering activities near a coastal region are effective in inducing rapid changes on coastal landforms resulting coastal hazards. Development activities, global warming, climate change and sea-level rise not only introduce any new types of coastal hazards, but they also stimulate the existing hazards. The southern coastal Tamil Nadu of India faces severe such threats due to rapid changes in geology and geomorphology, sea-level change, tropical cyclones and associated storm surges.

The ratio of people living in coastal zones compared with available coastal lands further indicates that there is a greater tendency for people to live in coastal areas than inland. According to the United Nations Environment Programme (UNEP) report, the average population density in the coastal zone was 77 people/km2 in 1990 and 87 people/ km2 in 2000, and a projected 99 people/km2 in 2010 [1]. Collectively, this is placing both growing demands on coastal resources as well as

Citation: Joevivek V, Saravanan S, Chandrasekar N (2013) Coastal Vulnerability and Shoreline Changes for Southern Tip of India-Remote Sensing and GIS Approach. J Earth Sci Clim Change 4:144. doi: 10.4172/2157-7617.1000144

increasing people's exposure to coastal hazards. At least, 200 million people were estimated to live in the coastal floodplains in 1990 (in the area inundated by a 1 in 1000 year flood), and it is likely that their number will increase to 600 million by the year 2100 [2]. Furthermore, global climate change and the threat of an accelerated sea-level rise exacerbate the already existing high risks of storm surges, severe waves, and tsunamis. Over the last 100 years, global sea level rose by 1.0-2.5 mm/y. Present estimates of the future sea-level rise induced by a climate-change range from20 to 86 cm for the year 2100, with a best estimate of 49 cm. It has been estimated that a 1-m rise in sea-level could displace nearly 7 million people from their homes in India [3]. Hegde and Reju [4] have used coastal vulnerability index to evaluate the hazardous zones at mangalore coast, India.

Scientific study of the natural hazards and coastal processes of the Indian coast has assumed greater significance after the December 2004 tsunami because the country learned lessons on the impact of natural hazards in terms of high damage potential for life, property, and the environment. The nation's rapidly growing population of coastal residents and their demand for reliable information regarding the vulnerability of coastal regions have created a need for classifying coastal lands and evaluating the hazard vulnerability.

The present study, therefore, is an attempt to develop a coastal vulnerability index (CVI) for the Southern tip of India using six relative risk variables. In this present work, shoreline changes and vulnerability level along the southern coastal Tamil Nadu have been assessed using remote sensing and GIS. The erosion and accretion made in different parts of the study area have been measured and analysed. The coastal vulnerability index (CVI) has been used to map the relative vulnerability of the study area and also characterize the vulnerability of the coast due to coastal processes and human activities such as understand the mining effect on coastal areas. Coastal sand mining is also actively pursued along the coast (Figures 1a and 1b). Human activities, particularly those that impound sand, can cause beaches to change. These include sand being obstructed from reaching the coast by damming rivers, as well as coastal engineering structures (for example, groins and jetties) that can trap sand moving along the coastline, depriving other areas from receiving sand. Anthropogenic activities like construction of building, sand dune destroyed for Tourism development and sand mining activities (Figures 2-4). Dense populations near by the coast (Figures 5a-5c), are also responsible for erosion, reduction in width of beaches.

STUDY AREA AND GEOMORPHOLOGY

The study area lies between Manavai and Vijayapathi, southern tip of India (Figure 6). The backshore of the beach is limited by urban infrastructures. The Mean Sea Level (MSL) is 2m above the 0m depth chart datum (CD).

Figure 1a: Sand mining near Kanyakumari coast.

Figure 1b: Sand mining near Kanyakumari coast.

Figure 2: Constructions along the Kanyakumari Coast.

Figure 3: Sand dunes levelled for tourism Sothavilai, Kanyakumari .

Figure 4: a) Exposed ground water in the sand mining operation near the coast. b) Total sand dune were removed with a view of the sea.

a b c

Figure 5 a, b, c: Coastal erosion due to dense population nearby coast.

The local mean tidal range is 0.5m and the maximum tidal level above CD is 1.60 m. The study area has minor river systems such as Palaiyar, Namiyar, Hanuman Nadhi and seasonal streams namely, Nilapparai channel and Puttanar channel. Figure 6 has shown the coastal geomorphology along the study area. Kanyakumari coast (Cape comorin) has major cliffs which are projected towards the sea forming a promontory. Pambar estuary adjoined with sea in the western part of study area, which is one of the minor estuary and manakudi estuary is a major estuary in the district. This is formed by the irrigation canal excess water during monsoon, and the water drained from the irrigational fields mixing with sea. A small creek adjoined in the east coast of the study area locally called as Vaikkal. AVM (Ananda Victoria Maharani) canal runs parallel to the entire coast and mixed with Pambar estuary in the west end of study area. AVM canal runs 100 m from the shoreline in Kotilpad in a width of 5 m. But in the west part of study area, AVM canal 300 m away and running very thin at a width of 1 m wide. The net longshore transport is directed to the north-east along Kanyakumari coast [5].

MATERIALS AND METHODS

Coastal vulnerability index is computed based on the six vulnerable parameters namely shoreline change rate, coastal slope, relative sea level change, mean wave height and mean tide range [6]. According to that, vulnerable parameters and its risky level is presented in Table 1. The process is mathematically described as below,

$$CVI = \sqrt{[a \times b \times c \times d \times e \times f / 6]} \quad (1)$$

Where, a=Geomorphology, b=Shoreline change rate (m/yr), c=Coastal slope (deg), d=Relative Sea level change (mm/yr), e=Mean wave height (m), f=Mean Tide range (m). Based on the table, calculated CVI values are divided into five classes to highlight different vulnerabilities. The CVI value <4 implies very low risk level, 4 to 6 implies low risk level, 6 to 8 implies moderate risk level, 8 to 10 implies high risk level and >10 implies very high risk level.

For instance, Manavai beach has medium cliff geomorphology. So it comes under rank 2. The beach has experienced average shoreline change rate is -2.6363 m/year. So the variable comes under rank 4. Similarly, other four variables attain the rank level 5, 1, 2 and 4 respectively. The vulnerable index is estimated by multiplication of these ranks values divided by six. The result we obtained is 7.3029 (moderate category) (Table 4).

The parameters such as coastal slope, relative sea level change, mean wave height and mean tide range are estimated by Toposheet and naval chart datum. The parameter shoreline change rate is derived by remote sensing techniques.

The Landsat mission provides long period of earth surface data for the purpose of scientific and commercial developments. The present research we used Landsat-1 MSS, Landsat TM5 and Landsat ETM+ to evaluate the annual shoreline change rate along the present study area. The details raw Landsat data, their sensor type, date of acquisition and specifications are listed in Table 2. In general, raw data need to be calibrated to obtain a closer result of the surface reflectance. Thus we followed Gyanesh et al. for radiometric calibration and atmospheric correction of Landsat data.

Direct shoreline digitization provides complicated results due to the presence of water saturated zones in the vicinity of land - water boundary. Therefore, images are classified into two classes namely, land and water bodies using support vector machine (SVM) classifier in ENVI 4.3 software package. The result showing 92% accuracy, however, shoreline near sea - land intercept makes complexity due to wave foams create high intensity value in the Landsat data. To rectify this, Normalized Difference Water Index (NDWI) is used identify real water bodies based on Bo-Cai Gao [7]. Classified raster images are imported into ArcGIS software and geo-rectified by ground control points. Survey of India topographic sheet 1967 (1:50000 scale) is a base map imported into ArcGIS 9.0 and digitized using 60 ground control points which is obtained through handheld GPS. The rectified images are projected in UTM zone 43 of northern hemisphere with WGS 84 datum [8-10].

Table 1: Ranking for Vulnerability variables and CVI index

Variables	Rank for risk Level#				
	1	2	3	4	5
Geomorphology (GEO)	Rocky cliffs	Medium cliffs	Low cliffs, Alluvial plains	Cobble Beaches, Estuary	Sand beaches, Salt marsh, Mud flats
Shoreline change rate (SHL) (m/yr)	>3.0	1.0 to 3.0	-1.0 to 1.0	-1.0 to -3.0	<-3.0
Coastal slope (SLO) (deg)	>4.5	4.0 to 4.5	3.5 to 4.0	3.0 to 3.5	<3
Relative Sea level change (SLR) (mm/yr)	<1.8	1.8 to 2.5	2.5 to 3.0	3.0 to 3.4	>3.4
Mean wave height (MWH) (m)	<0.30	0.30 to 0.60	0.60 to 0.90	0.90 to 1.20	>1.20
Mean Tide range (MTR) (m)	>6.0	4.0 to 6.0	2.0 to 4.0	1.0 to 2.0	<1.0

Table 2: Source and specification of Landsat imagery used in the study.

S. No	Path/ Row	Acq. Date	Sensor	No. of bands and Spectral range	Spatial resolution	File type	Data source
1	143/054	21/06/2006	ETM+	1,2,3,4,5,7: 0.45-2.35μm 6.1,6.2: 10.40:12.50 μm 8 (PAN): 0.52-0.90 μm	30m 60m 15m	Geo- Tiff	USGS
2	143/054	19/10/2000	ETM+				
3	143/054	11/03/1992	TM5	1,2,3,4,5,7: 0.45-2.35μm 6 : 10.40:12.50 μm	30m 120m	Geo- Tiff	USGS
4	153/054	08/02/1973	L-1 MSS	1,2,3,4: 0.499-0.989 μm	60m	Geo- Tiff	USGS

Table 3: Shoreline width changes in different periods.

Station No	Station Name	Lat (N)	Long (E)	Shoreline width change (m)			
				1973-1992	1992-2000	2000-2006	1973-2006
1	Manavai	8°5'7.16"	77°28'46°47"	-15	-6	-66	-87
2	Agastiswaram	8°4'35.91"	77°31'29.83"	-9	-3	-54	-66
3	Muhilankudiyuruppu	8°4'27.12"	77°32'9.92"	-6	-3	-66	-75
4	Cape comorin (Left)	8°4'24.90"	77°32'52.26"	-12	-9	-72	-93
5	Cape comorin	8°4'36.56"	77°33'0.90"	-9	-9	-75	-93
6	Chinna muttam	8°4'46.56"	77°33'1.21"	12	9	51	72
7	Elusattupettu	8°5'5.27"	77°33'14.31"	6	3	42	51
8	Vivekanandapuram	8°5'10.08"	77°32'25.60"	6	3	15	24
9	Visvanarayanapuram	8°5'33.32"	77°33'55.12"	6	6	24	36
10	Sundanparappu	8°5'43.30"	77°33'58.61"	-6	-3	-21	-30
11	Kalluvilai(Arokiapuram)	8°6'7.91"	77°33'24.74"	-3	-3	-30	-36
12	Lipuram	8°6'41.82"	77°33'24.95"	3	3	18	24
13	Lakshmipuram	8°6'59.36"	77°33'29.72"	-9	-3	-18	-30
14	Vattakottai (Fort)	8°7'26.00"	77°34'0.98"	-6	-9	-24	-39
15	Anjugramam	8°7'52.67"	77°34'27.84"	-12	-3	-30	-45
16	Kootapuli	8°8'36.91"	77°36'10.96"	-6	-3	-27	-36
17	Chettikulam	8°9'14.86"	77°38'20.49"	-9	-3	-45	-57
18	Koodankulam	8°9'48.21"	77°42'51.02"	-12	-6	-54	-69
19	Idindakarai (Left)	8°10'6.48"	77°44'7.05"	-21	-6	-75	-99
20	Idindakarai (Right)	8°10'39.08"	77°44'56.98"	-12	-9	-69	-90
21	Vijayapathi	8°11'0.89"	77°45'17.45"	-9	-12	-63	-84

(+) sign implies accretion and (-) sign implies erosion

Further, edges in the classified images are extracted using Sobal filters (Rafael C. Gonzalez and Richard E. Woods, 1999). The extracted shorelines of Landsat images are projected in Arcmap. To calibrate shoreline change rate, toposheet 1967 is fixed as base shoreline. Based on the baseline, transact lines drawn for 21 sites using shape measurement tool. Shoreline change and its corresponding duration is used to estimate shoreline change rate along the study area [12-17].

RESULTS

Shoreline changes

Shoreline changes are presented with erosion rate, because it is an important valuable parameter for coastal vulnerable detection along the southern tip of India. Table 3 summarizes rates of shoreline changes for 21 sites in the entire stretch, including both erosion and accretion values. In Table 3, positive sign implies accretion and negative sign implies erosion. Shoreline changes from 1973 to 2006, it is noted that, South and East zones are highly changed due to sediment transport, erosion, accretion and also impact of Tsunami (Figure 7a-7d).

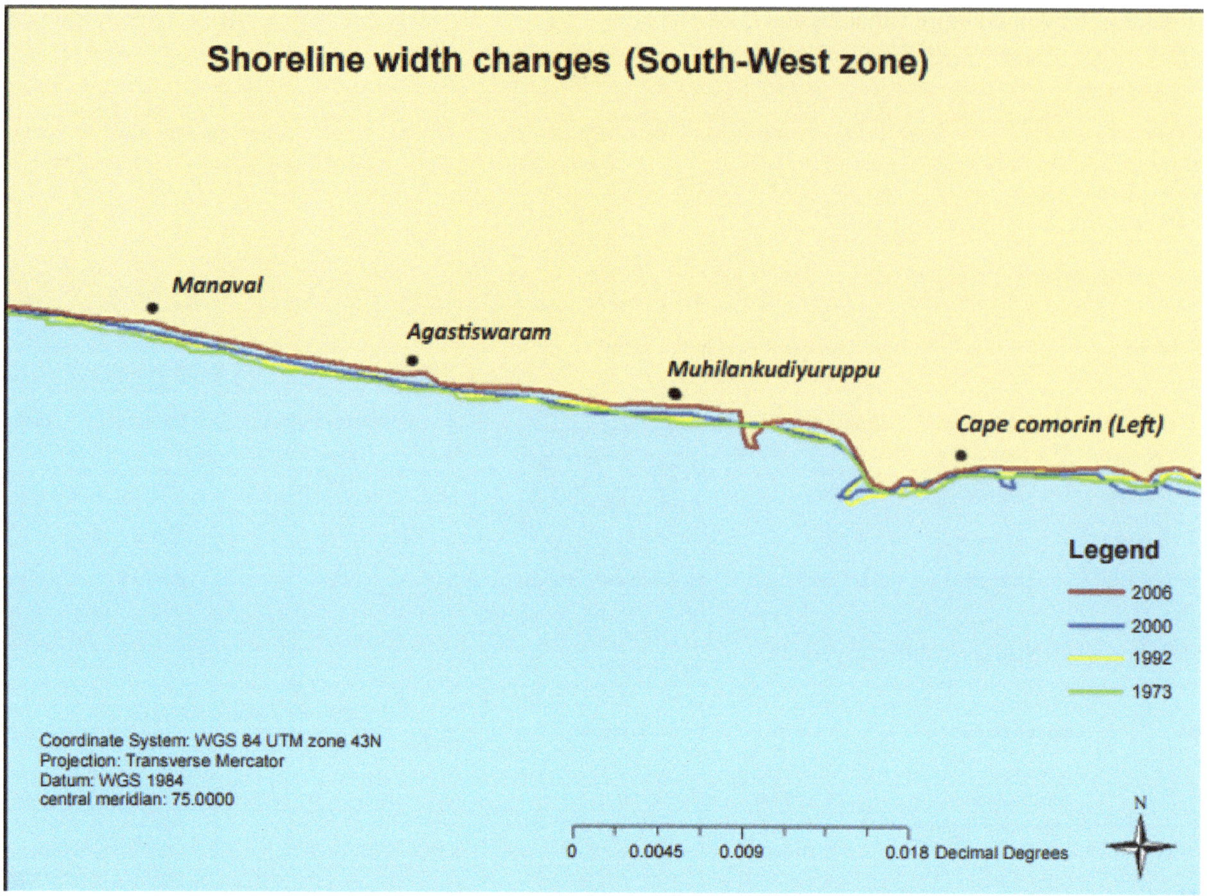

Figure 7a: Shoreline changes from 1973 to 2006.

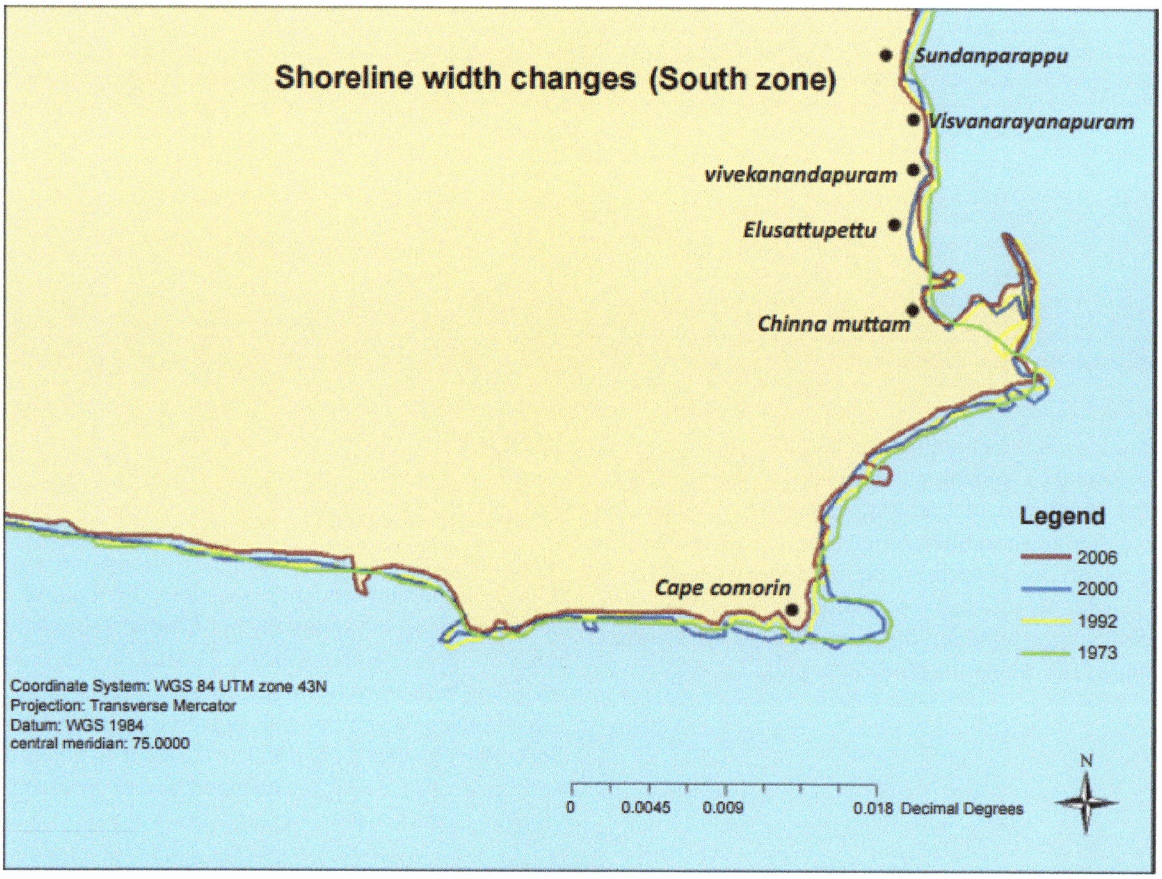

Figure 7b: Shoreline changes from 1973 to 2006

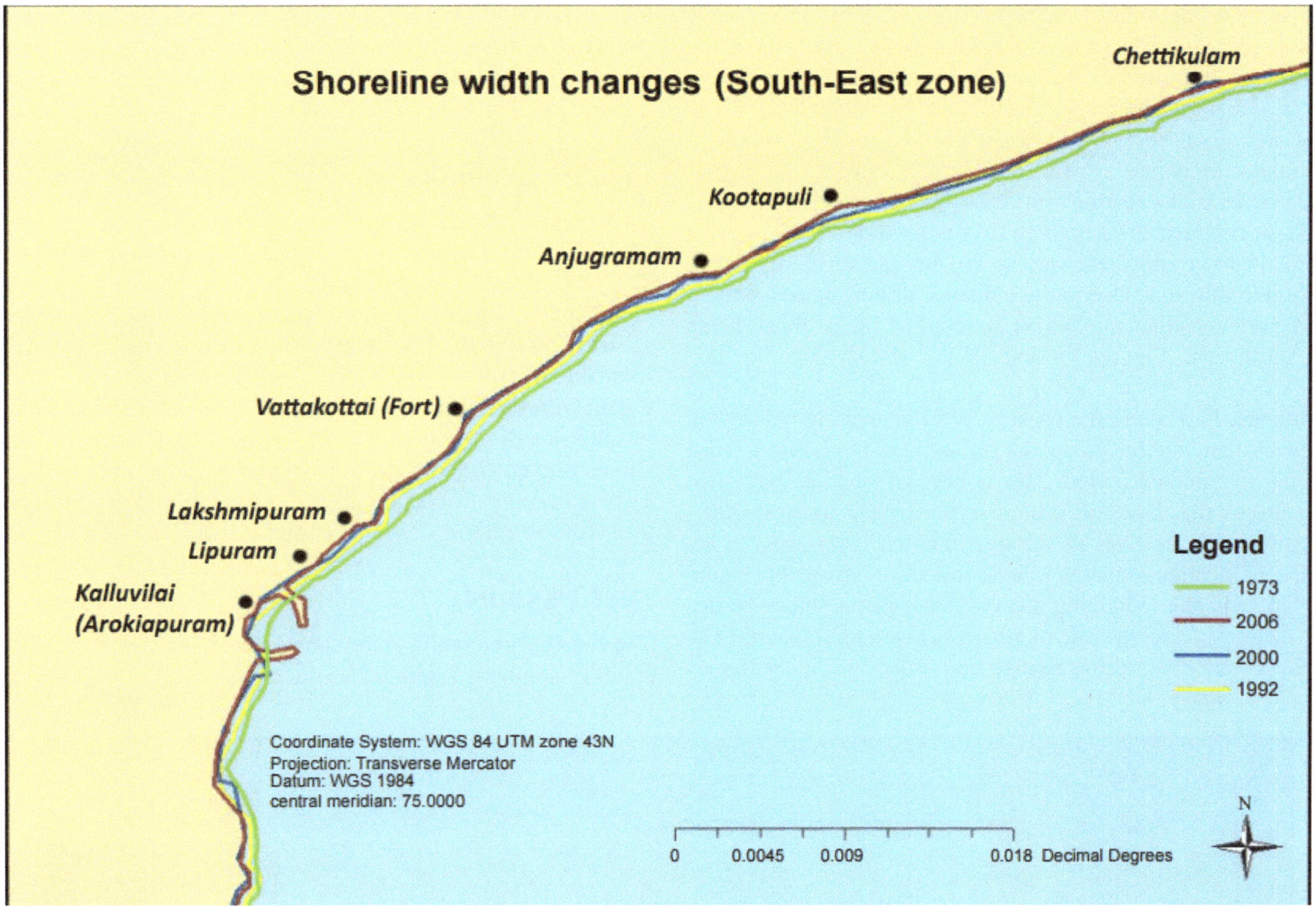

Figure 7c: Shoreline changes from 1973 to 2006.

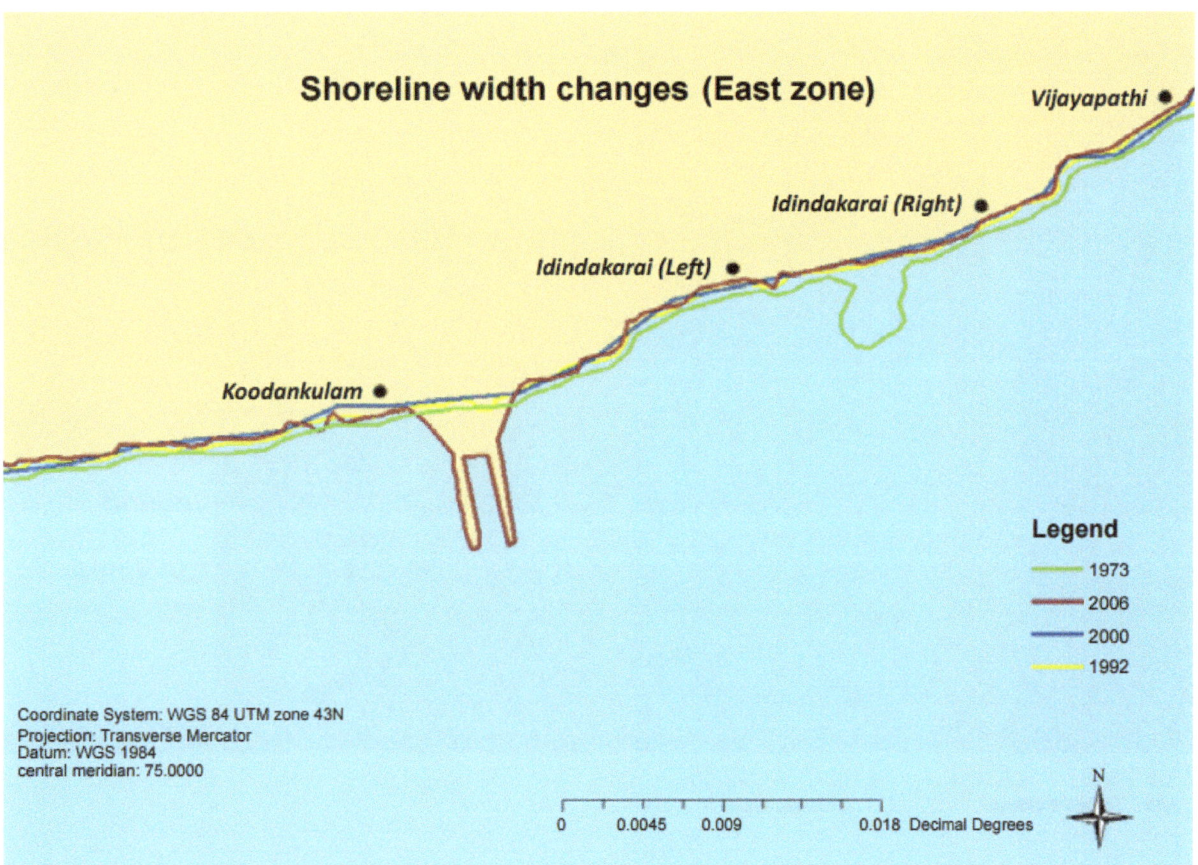

Figure 7d: Shoreline changes from 1973 to 2006.

As a result of the analysis, the most erosion changes were observed at the Cape Camoron, Idindakarai and Vijayapathi stations. In this area the net rate of erosion was found -4.6157 m/yr and no accretion trend observed for studied period in this area. Even though cape Camoron is a rocky cliff nature, it attains moderate erosion due to anthropogenic activities such as coastal construction, irrigation and tourism (Figures 1-3). In contrast, china muttam have achieved rate of accretion +3.8248 m/yr and there was no erosion detected for studied period in this area. Due to the influence of longshore sediment transport coastline movement have been likely from South to South-East. Figures 9a and 9b has shown that direction for erosion is North- Northeast and depositional direction in Northeast-East along the coast. It is an interesting nature that Kudankulam site have experienced erosion to accretion trend during the studied period. This is because, the site have two big groins that have been built in the period of 2000-2003. Unfortunately, it gives the eroding nature (down-drift) to the adjacent locations namely, Idindakarai and Vijayapathi (Figure 9) [15,16]. The remaining stations through the entire stretch have attained average rate of erosion as well as accretion. The most remarkable erosion change was found in Idindakarai. In the period of 1973 -1992, the site has experienced maximum coastline withdrawal of about 1.2 Km. The entire coast has attained maximum erosion at pre-monsoon and maximum deposition on monsoonal seasons [18-21].

Coastal vulnerability

With the reference of Table 3, coastal vulnerability zones were classified and mapped using ArcMap 9.0 software package. The coastal vulnerability map is shown in Figure 8. Coastal vulnerability category are classified based on the variables is shown in the Table 4. The results indicate that Koodankulam and Idindakarai stations were highly vulnerable zones. In contrast, Cape camoron and China muttam were low vulnerable zones. In other hands, a trade-off exists between Cape camoron and China muttam with respect to erosion and accretion processes. This is because of channel of sediment transport and littoral drift from south to eastern direction [23,24].

DISCUSSION

The coastal vulnerability map shows clear picture on hazardous status of southern Tamil Nadu coast.

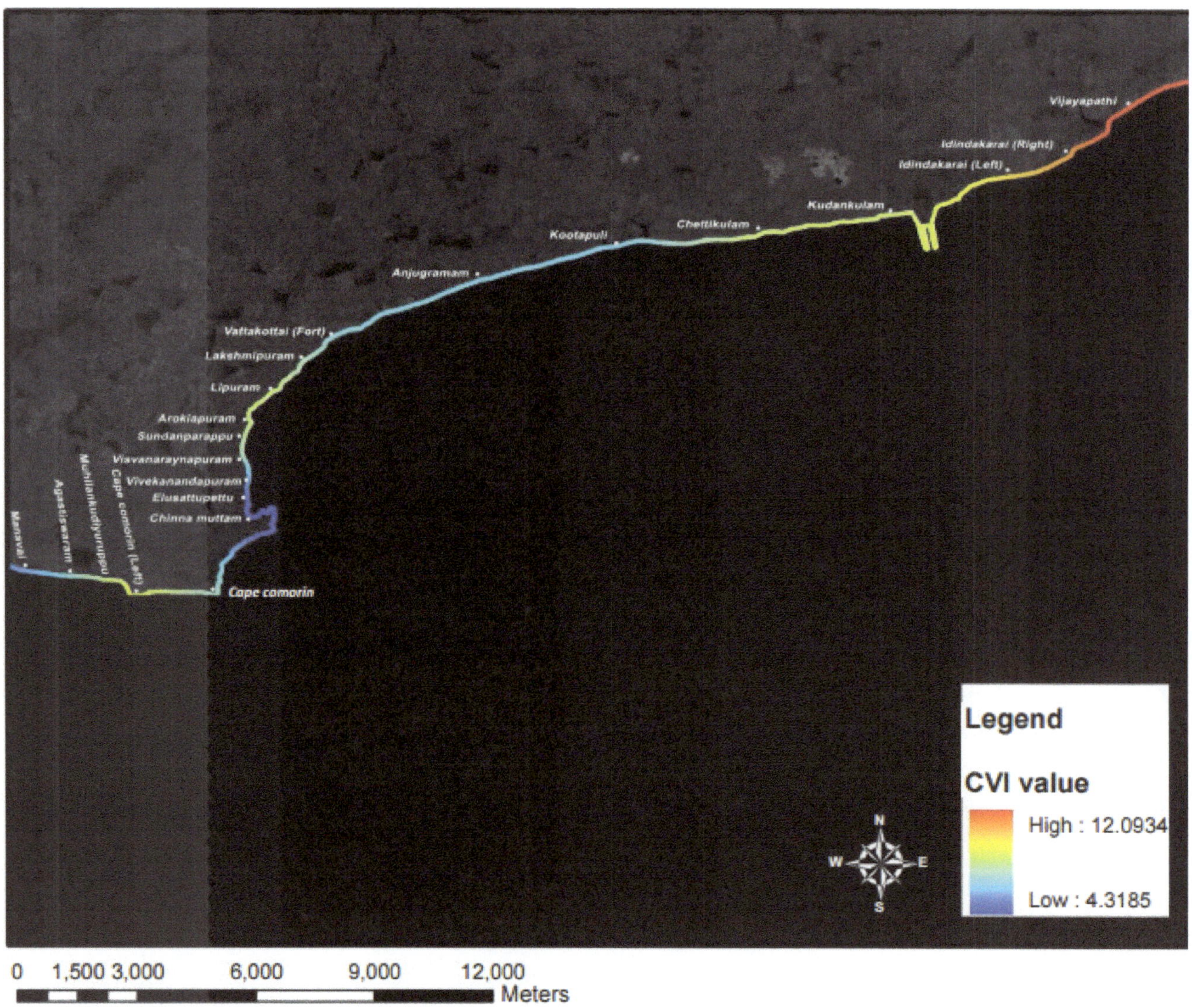

Figure 8: Coastal Vulnerability Index map.

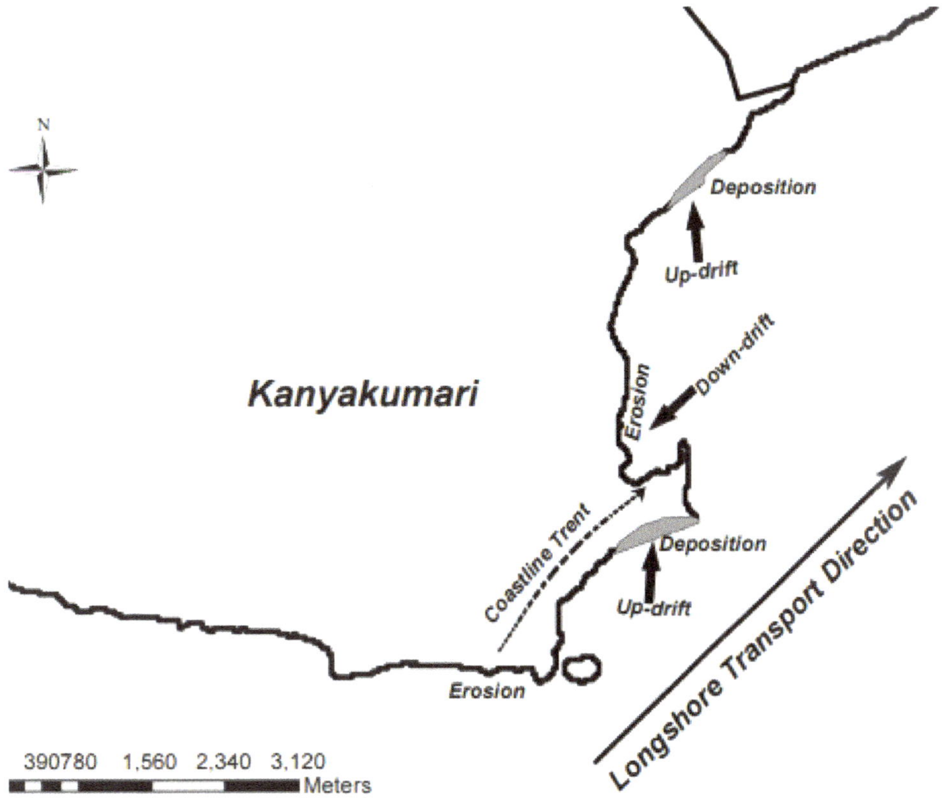

Figure 9a: Erosion and Deposition environment of the southern tip of Tamil Nadu.

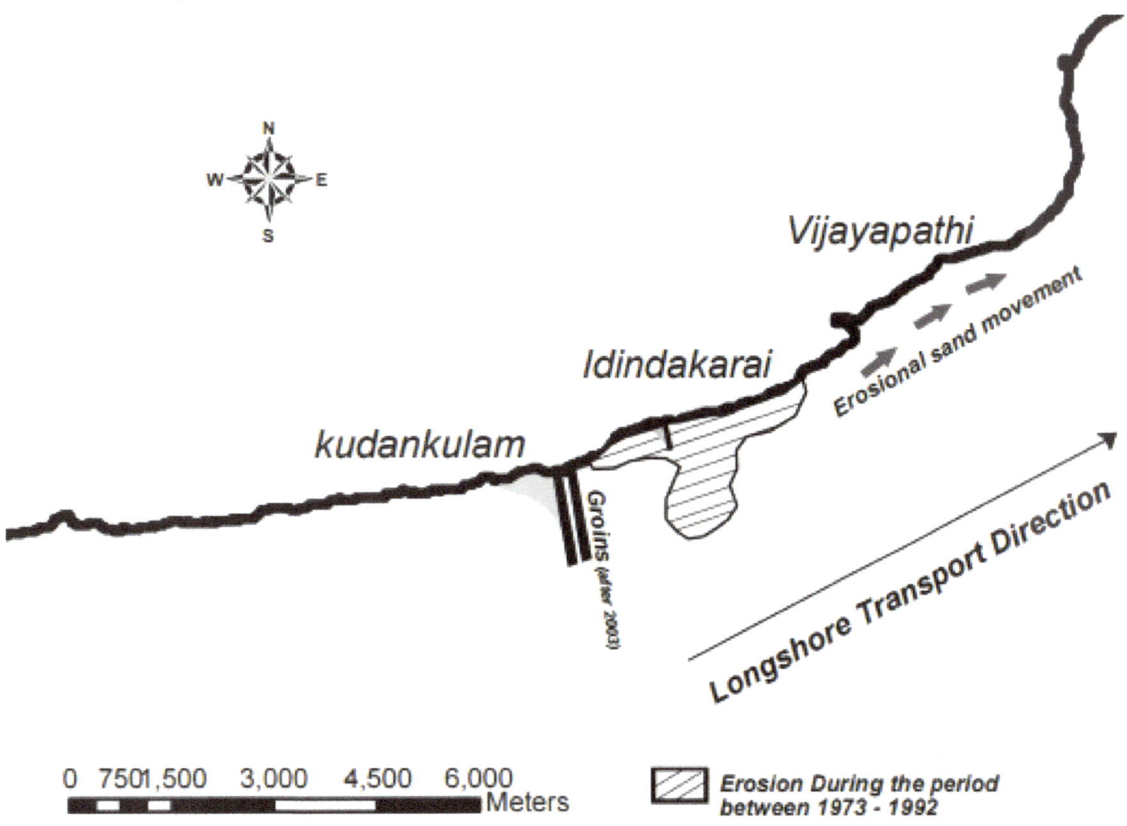

Figure 9b: Erosion and Deposition environment of the south-east coast of Tamil Nadu.

S. No	Location	Vulnerable Parameters						CVI value	Vulnerability Category	Remarks
		GEO	SHL	SLO	SLR	MWH	MTR			
1	Manavai	2	4	5	1	2	4	7.3029	Moderate	Sand sheet environment is highly dominated
2	Agastiswaram	2	3	3	1	2	3	4.2426	Low	All parameters are equally contributed
3	Muhilankudiyuruppu	3	3	3	1	2	3	5.1961	Low	All parameters are equally contributed
4	Cape comorin (Left)	4	4	4	1	2	4	9.2374	High	Erosion due to longshore sediment transport action
5	Cape comorin	4	3	4	1	1	2	6.9282	Moderate	Moderate erosion due to anthropogenic activity
6	Chinna muttam	3	1	3	1	2	3	3.0000	Very low	Rate of accumulation is high (up-drift)
7	Elusattupettu	3	3	2	1	2	3	4.2426	Low	Rate of erosion is high
8	Vivekanandapuram	3	4	2	1	2	3	4.8989	Low	Rate of erosion is high
9	Visvanarayanapuram	3	5	3	1	2	4	7.7459	Moderate	Very high erosion due to mud flats
10	Sundanparappu	2	5	4	1	2	4	7.3029	Moderate	Very high erosion due to less slope
11	Kalluvilai(Arokiapuram)	3	5	4	1	2	3	7.7459	Moderate	Erosion due to aquaculture
12	Lipuram	3	4	4	1	2	4	8.0000	High	Alluvial plain environment
13	Lakshmipuram	4	3	4	1	2	4	8.0000	High	Near estuary
14	Vattakottai (Fort)	2	4	3	1	3	4	6.9282	Moderate	Erosion due to coastal construction
15	Anjugramam	3	3	4	1	2	3	6.0000	Low	Parameters are equally contributed
16	Kootapuli	3	2	4	1	2	4	5.6568	Low	Less mean tidal range
17	Chettikulam	4	4	4	1	2	3	8.0000	Moderate	Cobble beach and estuary environment
18	Koodankulam	2	5	4	1	2	4	7.3029	Moderate	Moderate erosion and deposition due to groins formation
19	Idindakarai (Left)	4	5	3	1	2	4	8.9442	High	Erosion environment due to less altitude land surface
20	Idindakarai (Right)	4	5	3	1	3	4	10.9544	Very high	High erosion environment due to down-drift process
21	Vijayapathi	4	5	4	1	3	4	12.6491	Very high	Very high erosion due to estuary environment and down-drift process

Table 4: Vulnerable zone and significant remarks of each station along the study area.

For each of the sites, the union of each of the variables gave the CVI results as indicated in Table 4. About 50% of the computed values of CVI are 5 to 8, indicating a uniform vulnerability of most of the segments of the coast. The grid indicating the maximum vulnerability, i.e., 12.64, 12.24 and 9.23 correspond to the garnet sand mining region near the mouth of the estuary. The high value in CVI is the result of the high erosion rates experienced and the flat coastal slope accompanied by human settlements and sand mining activities. The next vulnerable region is the Lipuram, Kottapuli and Chettikulam, which has an index score of 7 to 8.0, just ahead of the average vulnerability of 4 to 6.5 (average of 8 values of CVIs obtained). This stretch exhibits a flat coastal slope with the densely populated in nearby coast of Kanyakumari coast. The lowest value of CVI of 3.46 and 1.57 covers the coastal stretch between Kanyakumari and Chinna muttom. This region exhibits rocky promontories and is moderately populated. Also the beaches may be considered to be in a state of dynamic equilibrium, exhibiting a cyclic phenomenon of erosion (during monsoon) and accretion (during summer). The presence of coastal structures such as piers and groins tend to alter the natural erosion and accretion processes along a beach. In addition, the presence of offshore features such as fringing and barrier reefs tend to absorb wave energy available for sediment movement.

It has been inferred that maximum hazard has occurred in the coast where there is a river mouth or an estuary as in the case of Lipuram, Lakshmipuram and Koodankulam. And minimum hazard has occurred in the coast where rock exposures are present in Southern Tamilnadu coast.

It has been inferred also that the impact of the coastal erosion/ storm/tsunami surge was high in the southern most part of the study area as most of the high vulnerable beaches falls on that region like Chinnamuttom, Kanyakumari, Manakudy and Pallam since they are awfully very much exposed to the refracted and diverted waves from Sri Lanka (Figure 10).

CONCLUSION

Without doubt, the best result for any CVI computation is heavily dependent on the quality of the data used and different types of data used, which influence the vulnerability of a particular coastal stretch between Manavai to Vijayapathi. With the available data, we find that the South tip of Indian coast is concerned in highly vulnerable category. Most of the segments (8 numbers) have a CVI value of range between 1.0 to 4.0 and they are taken to be moderate vulnerable due to rocky exposure and less population nearby coast. The quality of the present index can be enhanced by the addition of further variables, like wave height, tidal range, probability of storm, etc.

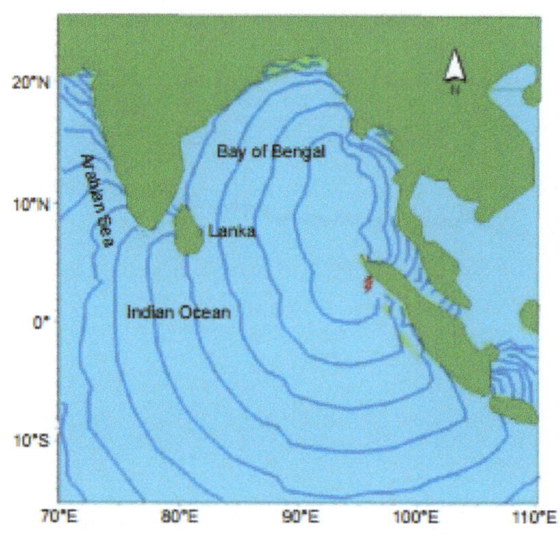

Figure 10: Refracted and diverted waves from Sri Lanka

The CVI is an indication of the relative vulnerability of the various segments of the southern tip of Indian coast. The map prepared for the Southern tip of Indian coast under this study can be used by state and district administrations involved in the disaster mitigation and management to take advance action to mitigate the effects of impending disasters and to prioritize areas for evacuation.

ACKNOWLEDGEMENT

The authors wish to thank the Department of Science and Technology, Government of India for providing financial support and necessary equipments in the form of project (ES/11/526/2000, dated: 09/12/2004) to perform this study. And also we wish to thank the Dr.Bhoop singh, Director, NRDMS Division, DST, New Delhi for their kind support to carrying out this study.

REFERENCES

1. UNEP (2007) Physical Alteration and Destruction of Habitats.
2. Mimura N, Nicholls RJ (1998) Regional issues raised by sea level rise and their policy implications. J Clim Res 11: 5-18.
3. IPCC (2001) IPCC Report, Working Group-I, Climate Change-2001: The Scientific Basis. Cambridge, UK: Cambridge University Press.
4. Hegde AV, Reju VR (2007) Development of coastal vulnerability index for Mangalore coast, India. J Coastal Res 23: 1106-1111.
5. Saravanan S, Chandrasekar N (2010) Potential littoral sediment transport along the coast of South Eastern Coast of India. Earth Sci Res J 14: 53-160.
6. Elizabeth A, Pendleton, Jeffress S, Robert W, Thieler E (2004) Coastal vulnerability assessment of assateague island national seashore (asis) to sea-level rise U.S. Geological Survey Open-File Report 2004-1020.
7. Bo-Cai G (1996) NDWI A Normalized Difference Water Index for Remote Sensing of Vegetation Liquid Water from Space. Remote Sens Environ 58 257-266.
8. Bayram B, Acar U, Seker D, Ari A (2008) A novel algorithm for coastline fitting through a case study over the Bosphorus. J Coastal Res 24: 983-991.
9. Ekercin S (2007) Coastline change assessment at the Aegean Sea coasts in Turkey using multitemporal Landsat imagery. J Coastal Res 23: 691-698.
10. Gornitz VM, Daniels RC, White TW, Birdwell KR (1994) The development of coastal vulnerability assessment database, Vulnerability to sea-level rise in the US south coast. J Coastal Res Spe Iss 12: 327-338.
11. Chander G, Markham BL, Helder Dl (2009) Summary of current radiometric calibration coefficients for Landsat MSS, TM, ETM+, and EO-1 ALI sensors. Remote Sens. Environ 113: 893-903.
12. Pethick JS, Crooks S (2000) Development of a coastal vulnerability index: a Geomorphological perspective. Environ Conser 27: 359-367.
13. Johnson DW (1919) Shore Processes and Shoreline Development. New York: Wiley.
14. Junchang J, David, Roy P (2008) The availability of cloud-free Land sat ETM+ data over the conterminous United States and globally. Remote Sens Environ 112: 1196-1211.
15. Kuleli T (2005) Change detection and assessment using multi temporal satellite image for North-East Mediterranean Coast. GIS Development Weekly.
16. Kuleli T (2009) Quantitative analysis of shoreline changes at the Mediterranean Coast in Turkey. Remote Sens Environ 167: 387-397.
17. Loos EA, Niemann KO (2002) Shoreline feature extraction from remotely sensed imagery. Geoscience and remote sensing symposium IGARSS IEEE Int 6: 3417-3419.
18. Roy D, Borak J, Devadiga S, Wolfe R, Zheng M, et al. (2002) The MODIS land product quality assessment approach. Remote Sens Environ 83: 62-67.
19. Ryu JH, Won JS, Min KD (2002) Waterline extraction from Landsat TM data in a tidal flat: a case study in Gosmo Bay Korea. Remote Sens Environ 83: 442-456
20. Sesli FA, Karsli F, Colkesen I, Akyol N (2008) Monitoring the changing position of coastlines using aerial and satellite image data: an example from the eastern coast of Trabzon, Turkey. Environ Monitor Assess 153: 391-403.
21. Shaw J, Taylor RB, Forbes DL, Ruz MH, Solomon S (1998) Sensitivity of the Canadian Coast to Sea-Level Rise: Geological Survey of Canada Bulletin. 505: 114.
22. Shepard FP (1963) Submarine Geology. New York: Harper and Row.
23. Thieler ER, Hammar-Klose ES (2000) National Assessment of Coastal Vulnerability to Sea-Level Rise: US Pacific Coast. US Geological Survey Open File Report.
24. Yamano H, Shimazaki H, Matsunaga T, Ishoda A, McClennen C, et al. (2006) Evaluation of various satellite sensors for waterline extraction in a coral reef environment: Majuro Atoll Marshall Islands. Geomorph 82: 398-411.

CHAPTER 13

THE IMPORTANCE OF ENVIRONMENTAL SAFETY EVALUATION OF BEACHES

Changchun Zhou

College of Geography and Tourism, Qufu Normal University, Rizhao 276826, Shandong, China

INTRODUCTION

60% of the world's population lives in areas within a distance of a 100 km from the sea at the end of the last century [1]. Coastal tourism is the largest market segment of global tourism [2-4]. At California beaches, each year nearly 150-400 million visits generated billions of dollars expenditures [5]. On the Miami Beach, more than 2 million visitors per annum brought over $130 million federal tax revenue [6]. At the Australian Gold Coast, 65, 000 to 120, 000 surfers spent AU$ 126-233 million per year [7]. Therefore, beach plays a significant role in tourism.

As a kind of tourism resource, beach can be viewed as an aggregate that is composed of the following environmental factors: sediment, topography, seawater, weather and climate, hydrology, human management, etc. In the activities of coastal tourism, some indexes of the coastal environment and management (such as seawater quality, hydrological and meteorological condition, sediment, safety education, danger warning, etc.) can have remarkable effects on tourists' experience, physical health, and even life safety. For example, it is estimated by a recent research that polluted coastal waters generate more than 120 million gastroenteritis (GI) episodes and 50 million Acute Respiratory Disease (ARD) episodes every year globally, causing $ 12 billion per year in public health costs [8]. For another instance, drowning is a major public health problem. Average 82 people suffered fatal drowning annually on Australia beaches [9] and the primary cause of death was rip currents [10]. Consequently, it is very important to evaluate environmental safety condition of beaches through an evaluation index system in order to guarantee tourism safety of beaches. Our objective is intended to help policy makers and beach managers understand and potentially improve the environmental safety condition of beaches, and help beachgoers better choose beaches and tourism activities which are fit for them.

There are many different beach rating systems in the world [7]. The Blue Flag maybe be the most well known of European award schemes and is run by the Foundation for Environmental Education (FEE), based in Denmark [11]. The Quality Coast Award introduced in 2007 by Environmental Campaigns which is an organization based in Wigan, UK. The Good Beach Guide is a book that is published by the Marine Conservation Society (MSC), a Non Governmental Organization (NGO) based at Ross on the Wales/England Border, UK. This guide consists of two main sections. The first section is water quality criteria and the second one provides a range of information such as safety, facilities, wildlife, seaside activities, etc. The Green Sea Initiative is supported by 30 public and private organizations such as local authorities, the private sector,

Citation: Zhou C (2016) The Importance of Environmental Safety Evaluation of Beaches. J Coast Zone Manag 19:416. doi:10.4172/2473-3350.1000416

statutory agencies, environmental and voluntary organizations. It is a major national project dedicated to the protection and improvement of coastal waters and beaches around Wales, UK and administered by Keep Wales Tidy. The Clean Beaches Council, a US Nonprofit Organization, launched 'Blue Wave Campaign' [12], and so on. These evaluation or rating systems pay close attention to the assessment of safety factors and thus play a key role in maintaining and guaranteeing the safety of beaches. Based on 821 questionnaires' survey in four bathing beaches of Wales from 1995 to 1996, 40% of tourists asserted that these rating activities had influence on their choices of bathing beaches. Since 2003, water quality, meteorology and hydrology condition have been monitored and predicted in the swimming seasons of the 23 primary bathing beaches of China. However, up to date, there is no corresponding evaluation or rating system of beaches at the national level in China. Though some scholars have done some work [13-24], the majority of their work mainly focuses on water quality evaluation. Wang [25] established an evaluation index system of environmental safety of beaches. Although his evaluation system is comparatively comprehensive, certain natural and management factors, such as tide, red tide, storm surge, tourist density, facility condition, personnel allocation, etc, are seldom included.

CONCLUSION

Therefore, at present, from the perspectives of providing health and safety services for beachgoers, the inclusion of these important natural and management factors in an evaluation index system of beaches is necessary to evaluate and classify beaches in China. And that's exactly what I am working on.

REFERENCES

1. Von Bodungen B, Turner RK (2001) Science and integrated coastal management. An introduction. Dahlem University Press, Berlin
2. Hall CM (2001) Trends in ocean and coastal tourism: the end of the last frontier? Ocean and coastal management 44: 601-618.
3. Honey M, Krantz D (2007) Global trends in coastal tourism. Center on Ecotourism and Sustainable Development.
4. United Nations Environment Program (UNEP) (2009) Sustainable Coastal Tourism. An integrated planning and management approach.
5. Given S, Pendleton L, Boehm AB (2006) Regional public health cost estimates of contaminated coastal waters: a case study of gastroenteritis at Southern California Beaches. Environmental Science and Technology 40: 4851-4858.
6. Houston JR (2002) The economic value of beaches: A 2002 Update. Shore and Beach 70: 9-12.
7. Williams A, and Micallef A (2009) Beach management: principles & practice. Earthscan, London.
8. Shuval H (2003) Estimating the global burden of thalassogenic diseases: human infectious diseases caused by wastewater pollution of the marine environment. Journal of Water Health 1: 53-64.
9. Surf Life Saving Australia (2008) 2008 National Coastal Safety Report. Surf Life Saving Australia, Sydney.
10. Sherker S, Williamson A, Hatfield J, Brander R, Hayen A (2010) Beachgoer's beliefs and behaviours in relation to beach flags and rip currents. Accident Analysis and Prevention 42: 1785-1804.
11. FEE (2013) Official web page for the Foundation for environmental education.
12. Mcleod WL, Evans C (2005) The blue wave campaign-creating and eco-ethic. Proceedings of the 14th biennial coastal zone conference, New Orleans, Louisiana.
13. Nelson C, Williams AT, Liu HB (1999) Award schemes and beach selection by tourists-a Welsh (UK) perspective. Coast Engineering 18: 156-168.
14. Ma CY (1997) Analysis and countermeasures of the status quality of the seawater of the sea baths areas along Qinhuangdao coast. Marine Environmental Science 16: 67-71.
15. Zheng JY, Qie ZY, Li XL (1998) The environmental quality assessment on the bathing beaches in south seashore of Qingdao City. Marine Environmental Science 17: 66-72.
16. Yang DN, Yuan DX, Xu PX, Deng YZ (1999) Monitoring and preservation of coastal bathing water quality of Xiamen in swimming season. Environmental Science 20: 75-78.
17. Chen CH, Wang L, Wang DR (1996) Water quality condition and analysis of its influencing mechanism of Xiuying bathing beach in Haikou Bay. ActaOceanologicaSinica 18: 45-52.
18. Han QQ, Zheng WH, Lian WL, Qiu JR, Weng F, Zhou M (2003) Investigation and analysis of water quality condition in Gulangyu bathing beach, Xiamen. Chinese Journal of Convalescent Medicine 12: 321-322.
19. Gu JQ, Liu GB, Zhang JM (1996) Effect of deposition parameters of development mode In Hebi beach. Geography and Territorial Research 12: 21-25.
20. Zhao NX, Yang DY (1996) Evaluation in physiological environment of a seaside tourism & vacation district. Marine Science 6: 66-68.
21. Lv Y, Zhang WD, Gao Y, Ning SX, Yang B (2011) Preliminary study on responses of marine nematode community to crude oil contamination in intertidal zone of Bathing Beach, Dalian. Marine Pollution Bulletin 62: 2700-2706.
22. Fan JF, Ming HX, Wu LJ, Liang YB, Li X (2011) Investigation of fecal coliform and typical enteric virus in representative beaches of China. Marine Science Bulletin 13: 72-81.
23. Zhang WW, Wang JY, Fan JF, Gao DL, Ju HY (2012) Effects of rainfall on microbial water quality on Qingdao No. 1 Bathing Beach, China. Marine Science Bulletin 66: 185-190.
24. Thoe W, Wong SHC, Choi KW, Lee JHW (2012) Daily prediction of marine beach water quality in Hong Kong. Journal of Hydro-environment Research 6: 164-180.
25. Wang XQ (2006) Research on evaluating environment safety of bathing beaches. Doctoral degree dissertation of China University of Geosciences: 158.

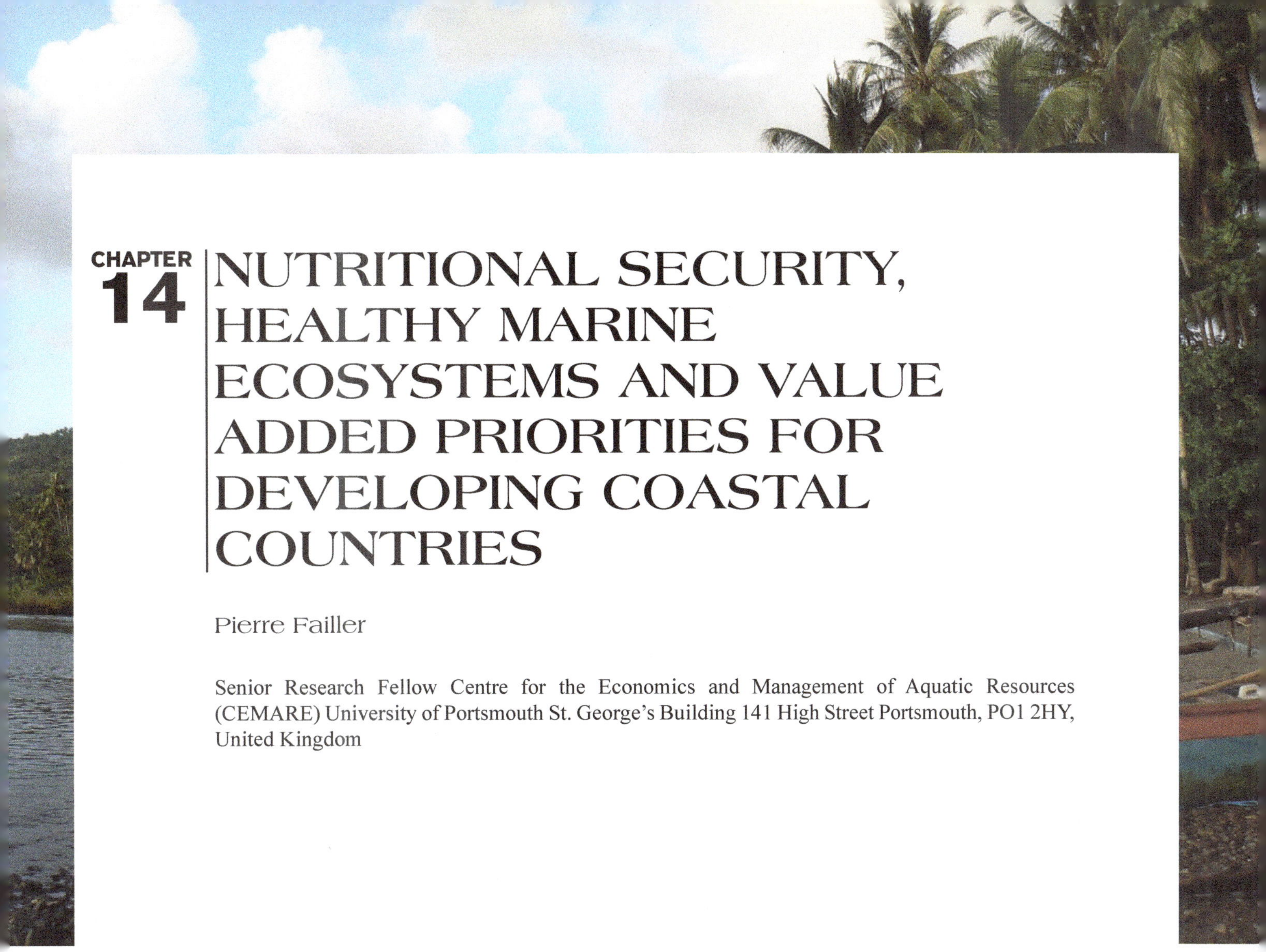

CHAPTER 14

NUTRITIONAL SECURITY, HEALTHY MARINE ECOSYSTEMS AND VALUE ADDED PRIORITIES FOR DEVELOPING COASTAL COUNTRIES

Pierre Failler

Senior Research Fellow Centre for the Economics and Management of Aquatic Resources (CEMARE) University of Portsmouth St. George's Building 141 High Street Portsmouth, PO1 2HY, United Kingdom

According to FAO estimates, global consumption of fish, currently at 140 million tonnes, is likely to reach 200 million tonnes by 20301 . Industrialised countries where households have strong purchasing power will increase their demand, while developing and emerging countries will use this as a basis to increase both their aquaculture production and the number of fish caught for export. Overall, the demand from international markets will lead to an increase of fish trade from low income countries to wealthy ones with severe nutritional consequences for population who rely heavily on fish for animal proteins. Considering this, International agencies such as FAO, UNEP, UNDP and UNIDO, in conjunction with regional and national institutions have to set up priorities to face this situation and contribute to the implementation of a more equitable fish trade. Three priorities can be identify. The first priority is to ensure a nutritional security for the population of developing countries that are exporting more and more fish. The poorest households still need to be able to continue to have access to animal protein from fish at affordable prices. Developing countries must therefore find a way to accommodate supplying both domestic and international markets. Exports must not place a strain on either the price or the range of goods offered to the local communities. The second priority is the stop the degradation of marine environment. Marine resources are either being over- or fully exploited in all developing countries. Due to the change in the structure of ecosystems brought about by climate change, southern countries must replenish fish stocks to ensure that marine ecosystems are as resilient as possible to exogenous shocks. As such, foreign vessels should be granted access to national waters only if there is proof of a surplus of fish stocks or the existence of a regional management fishery organisation. The third priority is the increase of the value addition of fishery products. Since the value of fish primarily depends on its natural qualities and size, the care taken in catching, handling, storing and transporting is an important element in distinguishing among top-quality, second-class and downgraded fish not suitable for export. For instance, the price difference per ton between top- 1 FAO (2012), The State of World Fisheries and Aquaculture. See http://www.fao.org/docrep/016/i2727e/i2727e00.htm quality and second-class fish is around €1,000, and about €3,000 for downgraded fish2 In Mauritania. The use of fishing methods that do little to foster quality results is estimated to lead to €80 million in lost of profits per year - almost equivalent to the sum of the country's yearly exports of fish. Unfortunately, this example is true in other developing countries. Thus, most of the wealth naturally generated by marine ecosystems is wasted through carelessness. While such waste had little impact twenty or thirty years ago, today it is increasingly damaging, both economically and environmentally. Marine ecosystems are under severe stress

Citation: Failler P (2014) Nutritional Security, Healthy Marine Ecosystems and Value Added Priorities for Developing Coastal Countries. J Fisheries Livest Prod 2:e108. doi:10.4172/2332-2608.1000e108

due to the tendency of large and small-scale fishing vessels to focus on quantity, at the expense of quality. Shifting to higher quality products would allow fishermen to generate their current turnover while greatly reducing their catch. There is a strong case, therefore, for developing countries to opt for commercial development based on quality products. For fresh, refrigerated and frozen fish, the care taken by operators is reflected in a higher selling price. For developed countries importers, this segment is not only more profitable, but also more promising for the future, as a result of the gradual shift in consumer preferences from canned to fresh quality fish3 (such as fish stored in minus 60 °C). In other words, adding value does not necessarily imply fish processing. Using the 'quality' criteria to determine access to fisheries resources holds promising prospects for conservation, particularly considering the failure of current fisheries management systems. Developing countries fisheries regulations, which have been modelled on those applied in developed countries, have often been ignored. The context is certainly different: the means for implementing policies in developing countries are scarce and enforcement capacities are limited. Furthermore, considering for instance the failures of the European Common Fisheries Policy, despite a €300 million annual budget, it is unrealistic to expect developing countries with much more meagre budgets to meet the same standards. However, it is possible to take advantage of the market's appetite for 'quality' and related criteria (for example, rules of origin, HACCP and ISO production process certification, fair trade, Code of Conduct for Responsible Fishing, etc) to improve fisheries management. This would entail regulating fisheries not by attempting to monitor access to marine resources but by ensuring that all catches comply with international quality standards. In practice, it would mean giving up catches on fish that have not reached the right sexual maturity, as well as eliminating practices like non-selective fishing, the use of gears that are damaging to fish, and poor preservation methods that lead to wastage (see www.esoctproject.org for a presentation of the novel method for assessing the full cost of fishing practices and fishery policy, from the environmental, social and economic point of view). Such a mechanism would be set up and maintained by the fishing industry itself, and would therefore provide greater accountability by the industry. This is especially important considering that, over the last thirty years, the industry has felt increasingly marginalised and excluded from any decision-making process that affects it, due to governments' obstinate insistence on setting up 2 Failler P. et al. (2006). Évaluation des stocks et aménagement des pêcheries de la ZEE mauritanienne. Rapport du 5ème Groupe de travail IMROP, Nouadhibou, Mauritanie, 7-17 décembre 2002, United Nations Organisation for Food and Agriculture, Rome, 2004, 203 p. 3 See on this topic: Failler et al. (2008). Future prospects for fish and fishery products; 4. Fish consumption in the European Union in 2015 and 2030 - Part 2. Country projections. FAO Fishery circular No. 972/4 Part 2, 400 p. centralised fisheries management systems. Attempts at initiating participative processes for fisheries management are indeed promising, but have a long way to go. Trading away fisheries resources to meet economic growth requirements is hardly conceivable today, especially if the ultimate objective of national development policies in developing countries is to reduce poverty. Rather, it is crucial to consider fisheries resources as a source of economic growth and social development, and in this regard, the health of marine ecosystems cannot not be jeopardised. Overall, this demonstrates the need for consensus-building between policy makers in the areas of fisheries and trade, and to a greater extent, those operating in the spheres of microeconomics, cooperation and development. Dialogue between these policy spheres should result in a consideration of the ways in which trade and fishing interact, as well as the specific objectives and constraints in each area (the need for exporting on the one side, and protecting resources on the other). This dialogue should then contribute to the development of strategies aimed at optimising the potential of marine resources. The emergence of links between fishing and trade policies also offers the merit of putting the environment at the forefront of the trade agenda. Ultimately, forging strong links between fishing and trade policies will contribute to the emergence of good governance in the fishing sectors.

REFERENCES

1. FAO (2012) The State of World Fisheries and Aquaculture.
2. Failler P, Diop M, Aliou Dia M, Abdallahi C, Tous P (2002) Évaluation des stocks et aménagement des pêcheries de la ZEE mauritanienne. Rapport du 5ème Groupe de travail IMROP, Nouadhibou, Mauritanie, Rome, Italy.
3. Failler P, Walle GV, Lecrivain N, Himbes A, Lewins R (2008) Future prospects for fish and fishery products: 4. Fish consumption in the European Union in 2015 and 2030 - Part 2. FAO publications, Rome, Italy.
4. www.esoctproject.org

CHAPTER 15

THE SCOPE OF MARINE SPATIAL PLANNING AND INTEGRATED COASTAL ZONE MANAGEMENT: NEW CHALLENGES FOR THE FUTURE

García-Sanabria Javier

Faculty of Marine and Environmental Sciences, University of Cádiz, Spain

"The sea, the great unifier, is man's only hope. Now, as never before, the old phrase has a literal meaning: we are all in the same boat."

Oceanographer: Jacques Yves Cousteau.

EDITORIAL

Marine environment is home to a rich biodiversity and long supported tradition of economic activities. Although we have used it for thousands of years, its importance to society, to life on Earth, as we know it today, goes far beyond what we imagined.

In this context, marine planning emerged in the seventies as a response to the interest of the States in the establishment of Marine Protected Areas [1]. The best example in this regard is the Australian

Great Barrier Reef Park. Today, the rapid growth that marine planning is undergoing is due to the interests of energy production [2], namely the rise of offshore-wind farms [3]. These interests are the main underlying reasons for the marine spatial planning efforts undertaken in regions as diverse as Germany, the UK, Belgium or the U.S.

Citation: Javier GS (2014) The Scope of Marine Spatial Planning and Integrated Coastal Zone Management: New Challenges for the Future. J Coast Dev 17:e109. doi: 10.4303/1410-5217.1000e109

With these motivations, international and national policies have increasingly changed towards promoting a framework for marine management oriented to the governance of these spaces [4]. Indeed, the

United Nations have been promoting since the 1992 Rio Conference the need for new management tools for maritime areas, establishing the objective of the parties to have management tools by 2002. In result, in the first decade of the XXI century there have been many advances in this regard; the UNEP Regional Seas Programmes have been particularly successful in this field, highlighting the success of the Mediterranean Sea Program. The European Union has responded to these advances developing an Integrated Maritime Policy which has established a suitable framework to manage marine areas of the

Member States. Other countries such as the USA, UK, Australia and Canada have also made significant improvements. Then, it seems clear that public policies begin to reflect changes that point towards governance as better management model for marine areas. But what does it means for managing marine-coastal areas?

Marine ecosystems, and the activities they support, have an intense relationship with ecosystems and activities located in the emerged part of the coast [5]. Indeed, Chapter 17 of Agenda 21 [6], states in its 17.39 Article, in relation to the protection of the marine environment, the following: public bodies for coastal-marine coordination should provide the capacity and authority needed to monitor all activities and sources of marine pollution located on land, and propose appropriate control measures.

These recognized relationships between land and sea environment are frecuently not translated into the management systems [7]. Two distinct management areas are distinguished. The first covers coastal terrestrial areas and coastal marine waters [8], where landsea relationships are very close and justify the approach to this scope of management through the perspective of integrated coastal zone management. Therefore, marine spatial planning, as also land planning, should be conceived as a management tool of this continuous land-sea area. The second management area would consist of marine waters, where ecosystems and activities do not keep such close relationships with the terrestrial area. In this case, the approach would be the integrated management of marine areas, in which maritime spatial planning would be a tool for management as well, but under this different and more international perspective [9].

In conclusion, in the next decades more efforts should be focused on bring clarity in how to organize current management systems for introducing the connectivity between land and sea as a way to improve the coastal and marine governance.

REFERENCES

1. Douvere F (2008) The importance of marine spatial planning in advancing ecosystem-based sea use management. Marine Policy 32: 762–771
2. Suárez de Vivero JL (dirección científica) (2001) Los Océanos. Medio Ambiente,
3. recursos y políticas marítimas. Ediciones del Serbal pp.310
4. Ehler CY Douvere F (2009) Marine Spatial Planning. A step-by-step approach toward Ecosystem-based Management. Intergovernmental Oeanographic Commision. Manual and Guides, nº 53, ICAM Dossier nº6. Paris, UNESCO pp.99
5. Suárez de Vivero J, Rodríguez Mateos JC, Florido del Corral D (2009) Geopolitical factors of maritime policies and marine spatial planning: state, regions, and geographical planning scope. Marine Policy 33: 624-634.
6. Barragán JM, Chica JA, Pérez ML (2011). Gestiónintegrada de áreaslitorales en España, Biodiversidad en España. Base de la sostenibilidad ante el cambio global. Observatorio de la Sostenibilidad en España. 2011: pp. 380-389, ISBN: 978-84-8476-433-5.
7. UNCED (1993). United Nations Conference on Environment and Development. Río de Janeiro (1992). Important results of UNCED were: Rio Declaration, Agenda 21, Convention on Biological Diversity, Framework Convention on Climate Change.
8. Barragán JM (2014). Política, gestión y litoral. Unanuevavisión de la gestiónintegrada de áreaslitorales. UNESCO y Editorial Tébar. 620 pp. ISBN. 978-984-7360-518-512.
9. Barragán Muñoz JM & Arenas Granados, Pedro (2012). ManejoCosteroIntegrado en Iberoamérica: Diagnóstico y propuestasparaunanuevapolíticapública. Red IBERMAR (CYTED), Cádiz, 152 pp.ISBN13: 978-84-695-1018-6.
10. García-Sanabria J (2014). La planificaciónespacial marina: unaherramientaútilparadiferentesámbitos de aplicación. Actas del VII CongresoInternacional de Ordenación del Territorio. 212-226.

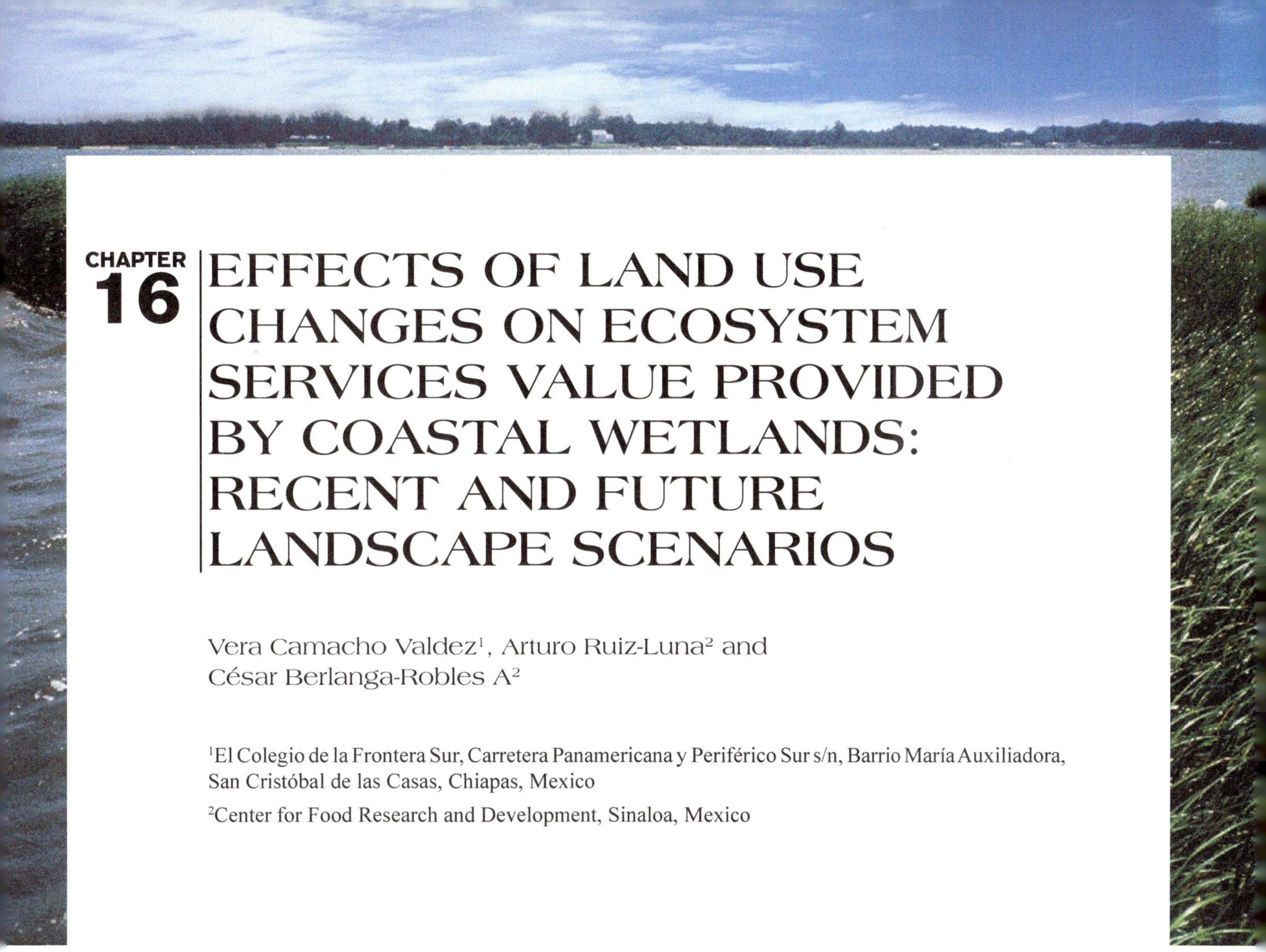

CHAPTER 16

EFFECTS OF LAND USE CHANGES ON ECOSYSTEM SERVICES VALUE PROVIDED BY COASTAL WETLANDS: RECENT AND FUTURE LANDSCAPE SCENARIOS

Vera Camacho Valdez[1], Arturo Ruiz-Luna[2] and César Berlanga-Robles A[2]

[1]El Colegio de la Frontera Sur, Carretera Panamericana y Periférico Sur s/n, Barrio María Auxiliadora, San Cristóbal de las Casas, Chiapas, Mexico

[2]Center for Food Research and Development, Sinaloa, Mexico

INTRODUCTION

The interest in the ecosystem services and the importance they have increased since the publication of the Millennium Ecosystem Assessment [1], with the report of global decline of 15 of the 25 listed ecosystem services, which anticipates a large and negative impact on future human welfare. Since then, one of the demands from the MA was to increase research on measuring, modeling and mapping ecosystem services, resulting in a positive response to increase the efforts to identify, quantify and value ecosystem services [2,3]. In fact, identifying the economic value of these services is essential in revealing their societal value, providing common metrics to facilitate comparisons among attributes and divergent scenarios in policy assessment [4]. These values differ significantly depending on the socioeconomic and geographic contexts, assigning higher values to wetland ecosystem services in countries with higher incomes, or valuing lower the hectare of a large wetland compared with the same but smaller wetland type [5].

The provisioning and quality of ecosystem services depend on the integrity and functioning of ecosystems, but the environment is systematically transformed, sometimes to increase production in agriculture, to increase space for human settlements or for resource extraction, with immediate or short time positive impacts on the local economy. However, it causes a decline in the provision of ecosystem services, which sooner or later will create a cost that must be paid by the population to compensate for the loss.

Although the impact of human activities is common to all natural ecosystems worldwide, coastal wetlands have been particularly affected because they receive the cumulated impact from all activities in the watershed. Although there are many drivers of change, Land Use/Land Cover (LULC) changes have been responsible, directly or indirectly, for the loss of 40% of all coastal wetlands and are the main source of pollutants discharged into coastal environments [6]. Moreover, in many parts of the world, local natural beauty, the availability of resources or the high productivity of natural wetlands are rapidly being converted into tourism facilities, salt evaporation or aquaculture ponds, housing developments, roads, ports, hotels and golf courses, thereby resulting in reduced or modified biodiversity, altered functional processes and the diminishing provision of ecosystem services to society, causing substantial social and environmental costs [7].

Mexico is not an exception to this global trend because the most important tourism infrastructure and technical agriculture have been developed along the coastal zone, modifying land use patterns and resulting in degradation and loss of natural ecosystems and their functions. Consequently, a decline in the

Citation: Valdez VC, Ruiz-Luna A, Berlanga-Robles AC (2016) Effects of Land Use Changes on Ecosystem Services Value Provided By Coastal Wetlands: Recent and Future Landscape Scenarios. J Coast Zone Manag 19:418. doi:10.4182/2473-3350.1000418

provision of ecosystem services to local communities occurs [8,9].

In northwest Mexico, the coastal zone of Sinaloa maintains ecologically important wetland zones with a high degree of naturalness, but particularly in southern Sinaloa, they are at risk mainly as consequence of governmental development plans to reactivate the regional economy and promote tourism and agriculture by creating facilities and infrastructure such as resorts, golf courses, hydrolectrical dams and irrigation and drainage channels.

Therefore, to guarantee the provision of ecosystem services from wetlands, policy interventions are needed to steer these processes and mitigate their negative impacts on ecosystems and society [10]. Such policy interventions require accurate assessments of the relevant ecological, socioeconomic, and political developments as well as of their potential future implications. LULC change analyses have been indicated as one of the high priority concerns for research and for the development of strategies for sustainable management [11]. In this context, it is important to produce information related to landscape changes and their effects on the value of ecosystem services provided by wetlands in this zone. This study focuses on landscape changes, but it also contributes an update to the current status and trends of wetlands, making it possible to compare and evaluate the output offered by alternative investments and highlighting the importance of wetlands as natural capital. Therefore, the main objectives of this study were to assess the dynamics of LULC based on recent Earth Observation data and forecast future changes in Ecosystem Service Values (ESV) provided by wetlands by modeling future LULC scenarios.

Study area

The Southern Coastal Zone of Sinaloa (SCZS) is integrated by four municipalities (Mazatlan, Concordia, Rosario and Escuinapa, from north to south) on the west coast of Mexico, with an approximately 120 km coastline and a total area of approximately 2500 km2 (Figure 1).

Despite that Rosario does not have coastline, the municipalities of the SCZS share similar environmental conditions, including diverse coastal wetlands such as mangroves, saltmarshes, estuarine and fluvial systems. The climate is warm and humid, with rainfall in the summer (928 to 1,457 mm) and an annual average temperature varying between 22 and 26°C [12].

The most important wetland ecosystems are represented by the Estero de Urias and the Presidio River at Mazatlan, the Huizache- Caimanero lagoon system and the Baluarte River in Rosario, and diverse lagoons and saltmarsh ecosystems integrated into the Marismas Nacionales (Sinaloa) system, which also includes the Teacapan Estuary in Escuinapa (Figure 1).

There are more than 35 urban and rural communities in the region, and all of these communities have less than 5,000 inhabitants, with the exception of Mazatlan, Villa Union, El Rosario, and Escuinapa (all connected by federal highways). Among them, Mazatlan is the most important city, with a population above 430,000 people, 87% of the total municipal population, and infrastructure to support tourism, industrial and fishing harbor, and other economic activities [12].

The population growth in Mazatlan (5% annual rate) and a population density of 173 people per km2 produces high pressure on natural land cover, and, because of the lack of appropriate land use planning and measures for sustainable development, environmental stress is also growing. The remaining municipalities have population densities below the state average (48 people per km2), but productive developments (aquaculture, agriculture, tourism) are growing, which also adds pressure to the local natural resources.

Despite the importance of coastal ecosystems and after several attempts, there are currently no natural protected areas in the region, except for the federally protected oceanic islands and mangrove systems.

METHODS

To analyze LULC changes and forecast trends for the next two decades in both land cover and ecosystem service values, the methodological approach used here included three steps: (i) the production of LULC thematic maps from 2000 and 2010; (ii) an estimation of change trends using Markov chains; and (iii) an analysis of the annual ESV flow based on wetland type and change tendencies.

LULC classification

In this study, the land cover data sets were produced from multi-spectral Landsat Thematic Mapper (TM) imagery (path/row: 31/44), acquired from the USGS Global Visualization Viewer (http://glovis.usgs.gov/), corresponding to the dry season in May 2000 and March 2010. Both images were clear and nearly free of clouds, previously projected to UTM coordinates (zone 13Q, WGS84) before limit the study area to the boundaries of municipalities and the physiographic province of the Pacific Coastal Plain (PCP) by a masking process. The image processing, GIS development and output of the final coverage maps were conducted using IDRISI Taiga software and ESRI ArcGis 9.3.

Each image was updated with vector features such as rural and urban areas, channels, rivers and shrimp farms, which were digitized on-screen from QuickBird images from 2002-2010 available from Google Earth. These features were added as a mask in every spectral band, eliminating the corresponding pixels for further classification processes. The modified images were later independently classified into six informational classes of natural wetlands as defined by Berlanga et al. [13] using a supervised method with the maximum likelihood algorithm. Once the classification process was concluded, the previously defined polygon vectors were added and reclassified to produce a total of eight LULC categories (Table 1).

Table 1: Definitions of LULC types in southern Sinaloa.

ID	Type	Description
1	Littoral	Intertidal marine wetland
2	Coastal lagoon	Subtidal estuarine wetland
3	Saltmarsh class unconsolidated bottom	Intertidal estuarine wetland: intermediate floodplains by tidal action including land without vegetation

4	Saltmarsh class/ forested mangrove	Forested-shrub estuarine wetland: plant association formed by one or a combination of the four species of mangrove
5	Riverine	Permanentriverinewetland
6	Aquacultureponds	Human-made wetland: ponds for shrimp
7	Urban	Land used for towns and villages
8	Agriculture	Agricultural fields
9	Landcover	Land cover: tropical forests, secondary succession.

The validation of the output maps was conducted based on the proposal by Pontius et al. [14]. Each classified map was compared with a reference map to generate an observed sample matrix (cross-tabulation) that represents the absolute frequencies nij of the i-th category in the classification map (rows) and the j-th category in the reference map (columns). The values in the diagonal njj correspond to concordances in both the rows and columns classes. From this, a population matrix representing the proportions of the different categories in the classified map and the reference map (pij) was created:

$$p_{ij} = \left(\frac{n_{ij}}{\sum_j n_{ij}}\right)\left(\frac{N_i}{\sum_i N_i}\right)$$

where N_i is the observed frequency of the i-th category in the classified map (total of the i-th row).

The population matrix allows the estimation of unbiased statistics, as the quantity disagreement (Q) and allocation disagreement (A) parameters, calculated for each class and overall. The quantity disagreement parameter is derived from the differences between the proportions of categories in the classified map and the reference map, taking value of zero when the categories have the same proportion in both maps. The allocation disagreement parameter occurs when differences in the spatial allocation of the categories in both maps are present, and its value depends on the number of pixels in the classified map that need to be reallocated to increase the agreement with the reference map [14].

In addition to the Q and A disagreement values, the standard Kappa index (K') was also estimated. This index accounts for the expected agreement due to random spatial relocation of the categories in the classified map given the proportions of the categories in both maps, regardless of the size of the quantity disagreement parameter. The index takes values in the range from 0 to 1, when the observed agreement is equivalent to the statistically expected random agreement and when the agreement is perfect, respectively. All formulas of these indicators of agreement and disagreement are fully explained in [14].

Modelling LULC changes

This process was achieved using Markov chains to represent a spatial transition model, which is widely used to model LULC changes [15]. The Markov chain analysis follows a stochastic process in which distinct LULC categories are considered as states of the chain. A chain is defined as a stochastic process having the property that the value of the process at time t, Xt, depends only on its value at time t-1, Xt-1, and not on the sequence of values Xt-2, Xt-3,…,X0, which the process passed to arrive to the stage Xt-1 [16,17]. The Markov chain can be expressed as:

$$P\{X_t = a_j | X_0 = a_0, X_1 = a_1, \ldots, X_{t-1} = a_i\} = P\{X_t = a_j | X_{t-1} = ai\} = P_{ij.}$$

where the conditional probabilities Pij represents single-step transition probabilities or simply transition probabilities of the Markov chain; and Pij represents the probability estimated during the transition process from state i to state j in a given period of time [16]. Pij can be estimated from the values observed in a change detection matrix by tabulating the observed area of data i and j (nij) and dividing it by the sum of the number of times (Ni) that state i has occurred [17]:

$$P_{ij} = \frac{n_{ij}}{N_i}$$

The 2000 and 2010 land-cover maps were contrasted with a crosstabulation change detection matrix, and the statistical independence of LULC was tested with the chi-squared (X2) test [17]. Later, the transition probabilities Pij were calculated to output a single step transition probability matrix, a discrete-time Markov chain (DTMC), and this was generalized to an n-step transition probability matrix with the Chapman-Kolmogorov for Markov chains [16]:

$$P_{ij}^{n} = \sum_k p_{ik}^{l} P_{kj}^{(n-l)} \quad \text{for } 0 < l < n$$

In matrix notation:

$$P^{(n)} = P^{(l)} P^{(n-1)}$$

The matrix of n-step transition probabilities is obtained by multiplying the matrix of one-step transition probabilities by itself (n – 1) times P(n) = Pn [16]. We projected the matrix P for 2020 (n = 1) and 2030 (n = 2). The analysis was completed using the Idrisi Taiga Markov module [18].

Estimation of ESV

A wide array of market and non-market methods has been used to economically estimate the ecosystem services values of wetlands [19-22]. Based on the data set provided by Ghermandi et al. [23], which uses value references from empirical studies, Valdez et al. [24] applied a value transfer method to derive economic values for ecosystem services provided by coastal wetlands in Sinaloa, validating the method through a meta-regression model. Here, those values were used to obtain estimates of ESV for 2000, 2010 and for the 2020-2030 projection.

The ESVs for each land cover type and the total flow for the two study years (2000 and 2010) were calculated by multiplying the value coefficients obtained from Valdez et al. [24] by the respective cover extent. The change in the ESV was estimated by summarizing the value of each LULC type from 2000 and by calculating how each type changed in the following decade (2010). Based on the projected areas of the Markov chains model (ha), we calculated the total flow in the horizon years (2020 and 2030) using the same value coefficients as in previous years (2000 and 2010) with regard to the natural wetland categories.

RESULTS

Overall trend in the LULC

Once the imagery classification processes were completed, the accuracy of the output maps was validated to define the extent of each land cover category. From the confusion matrix, the proportion of correctly classified pixels was estimated at 0.83 (83%). Consequently the total disagreement was 0.17, and most of this, around 80%, corresponded to Q, while A rise up to 20%. This implies that the classification errors are mainly associated with the overestimation/underestimation of the areas of each category, rather than the distribution of the categories on the map. The classes with major allocation disagreement were Land cover (Lco) and Agriculture (Agc) reaching 97% and 100%, respectively (Figure 2). Conversely, the Littoral category (Lit) was classified without error and the disagreement in the Coastal lagoon category (Cla) was mostly due to A (>60%). Considering all the land use and cover categories, the standard Kappa index was K'= 0.77, which corresponds to a substantial agreement on the [25] classification scale.

Based on those results, the output maps were assumed to be a reliable interpretation of the landscape of the study area in both years analyzed. The LULC maps for the dry season in years 2000 and 2010 are displayed in Figure 3. The dominant covers for both dates were land cover (Lco) and agriculture (Agc), but regarding wetlands, as defined by Berlanga et al. [13] in their classification system, the most important were coastal lagoon (Cla), saltmarsh unconsolidated bottom (Sun) and saltmarsh forested mangrove (Sme).

Most of the cover classes displayed surface gains and losses, reflecting an spatial dynamics (Figure 4a), but at the final balance, saltmarsh/unconsolidated bottom and coastal lagoon, were the wetland covers that most positively contributed to the landscape transformation, with net gains of 1990 ha and 1718 ha, respectively, while saltmarsh/ forested mangrove and littoral classes had a net loss of 1567 and 529 ha (Figure 4b). The agriculture cover decreased considerably, losing a net area of 8,634 ha during the study period.

Table 2 shows the LULC change matrix from 2000 to 2010, and it is clear that significant change (13% of the total area) happened during the 10-year period, with the main changes attributed to interactions between agriculture soils and other land uses or coverages (Lco). Also the natural wetland classes (Lit, Cla, Sun and Smn) contributed to the Lco increasing with around 3500 ha or about 30% of the total gain for this cover from 2000 to 2010.

Table 2: LULC change detection matrix (ha) for 2000-2010.

2000											Total 2010
2010	Lit	Cla	Sun	Smn	Riv	Aps	Urb	Agc	Lco		
	Lit	**1,090**	7	1	0	0	0	0	29	25	1,152
	Cla	2	**17,189**	4,395	638	0	6	0	286	3	22,519
	Sun	1	2,453	**13,420**	2,665	0	29	0	3,354	16	21,938
	Smn	1	566	153	**7,270**	0	18	0	1,731	63	9,802
	Riv	0	4	0	6	**978**	0	0	14	4	1,006
	Aps	0	90	271	5	0	**2,208**	0	120	31	2,725
	Urb	0	0	0	0	0	0	**8,166**	706	1	8,873
	Agc	33	8	60	49	0	9	0	**40,389**	11,703	52,251
	Lco	553	484	1649	734	0	153	0	20,354	**111,120**	135,047
	Total 2000	1,680	20,801	19,949	11,367	978	2,423	8,166	60,885	129,063	255,313
	Change (ha)	-528	1,718	1,989	-1,565	28	302	707	-8,634	5,985	33,650
	Change (%)	-31.5	8.3	10	-13.8	2.9	12.5	8.7	-14.1	4.7	12.6

Although the highest proportion of change involved non-wetland inland covers, littoral and the only forested wetland class (Smn) suffered a considerable decrease of about 31% and 14% respectively. By contrast, coastal lagoon and saltmarsh/unconsolidated bottom classes increased about a tenth of their former area, while the artificial wetland Aps also gained around 300 ha.

LULC-Markov analysis

The simple-step transition probability matrix P, resulting from the contrast of the 2000 and 2010 study area LULC patterns, is shown in Table 3, displaying statistical dependence, X2 = 1.3 × 106 (P < 0.05). The resistance to change from one class to another can be observed on the diagonal of the matrix, representing the fraction of pixels that maintain the same LULC in the initial and final images. In this time period, littoral and saltmarsh (both unconsolidated bottom and forested mangrove) were the wetland classes most susceptible to change, together with agriculture. By contrast, coastal lagoon, riverine and aquaculture pond wetlands, together with urban category, were the most resistant to change. It highlights that lower resistance was observed among natural wetlands.

Table 3: Estimates of LULC transitional probabilities (2000-2010).

Class										
Lit	**Cla**	**Sun**	**Smn**	**Riv**	**Aps**	**Urb**	**Agc**	**Lco**		
Class	**Lit**	**0.65**	0	0	0	0	0	0	0	0
	Cla	0	**0.83**	0.22	0.06	0	0	0	0	0
	Sun	0	0.12	**0.67**	0.23	0	0.01	0	0.05	0
	Smn	0	0.03	0.01	**0.64**	0	0.01	0	0.03	0
	Riv	0	0	0	0	**1**	0	0	0	0
	Aps	0	0	0.01	0	0	**0.91**	0	0	0
	Urb	0	0	0	0	0	0	**1**	0.01	0
	Agc	0.02	0	0	0	0	0	0	**0.6**	0.1
	Lco	0.33	0.02	0.08	0.06	0	0.06	0	0.3	**0.9**

According to Table 3, the Lco class, involving different land use and coverages, was the most transformed, gaining surface from agriculture and littoral classes (P>0.30). Just regarding wetlands, saltmarsh/ forested mangrove class was the most affected among wetland classes, being transformed into saltmarsh/unconsolidated bottom class, with a transformation probability of 0.23, the highest if littoral to land cover change is disregarded; at the same time, the saltmarsh/unconsolidated bottom class decreased and transformed into the coastal lagoon class and the saltmarsh/forested mangrove class with probabilities of 0.22 and 0.01, respectively. The littoral class is also transformed into land cover with a probability of 0.33.

Extent of LULC categories

Based in the calculated transitional probabilities, the most extensive wetland categories during the study years (2000, 2010 and the predicted 2020- 2030) are represented by the coastal lagoon and saltmarsh (both unconsolidated bottom and forested mangrove) classes, amounting together around 21% of the area, increasing 1.1% in the whole study period. The land cover class, which represents the natural vegetation and other land cover classes, was always above 50% of the total study area, excepting in 2000 when occupied 48.1%. In addition, it is expected that for 2030, the agriculture class will loss an equivalent of 10% of the total area, while urban cover and aquaculture ponds will increase less than 1% of the same extent, probably growing on agriculture land.

All the categories will maintain their trend observed for the period 2000-2010, with the exception of the saltmarsh/ unconsolidated bottom class that maintained an increasing trend, but at the end it is expected a marginal reduction (193ha) during the 2030 projected year (Table 4).

Table 4: Total estimated area (ha) of each LULC in 2000, 2010 and the predicted years 2020 and 2030.

Land cover type	**2000**		**2010**		**2020**		**2030**	
	Area (ha)	**%**	**Area (ha)**	**%**	**Area (ha)**	**%**	**Area (ha)**	**%**
Lit	1,680	0.6	1,152	0.5	805	0.3	579	0.2
Cla	20,801	8.2	22,519	8.8	24,117	9.5	25,433	10
Sun	19,949	7.8	21,938	8.6	21,980	8.6	21,787	8.5
Smn	11,367	4.5	9,802	3.8	8,311	3.3	7,348	2.9
Riv	978	0.4	1,006	0.4	1,029	0.4	1,052	0.4
Aps	2,423	1.0	2,725	1.1	2,993	1.2	3,247	1.3
Urb	8,166	3.2	8,873	3.5	9,328	3.7	9,782	3.8
Agc	66,983	26.2	52,251	20.5	48,792	19.1	47,342	18.5
Lco	122,966	48.1	135,047	52.8	137,958	54.0	138,743	54.3
Totals	255,313	100.0	255,313	100.0	255,313	100.0	255,313	100.0

Variation of ESV

The estimation of the total Environmental Services Value (ESV) flow was obtained based only in the temporal patterns associated with natural wetland cover types (Table 5). It was calculated an increase from about $223 million (2007 USD dollars) up to $241 million when the ESV was projected to 2020. However, in the next projected decade (2030), the total flow decreases by approximately 2 million (2007 USD). The ESV of the saltmarsh classes (unconsolidated bottom and forested mangrove) accounts for 90% of the total estimated ESV, followed by the coastal lagoon (7%), and riverine (1%) classes. The total ecosystem service value increases over time, amounting to approximately 18 million (2007 USD).

Table 5: Temporal patterns of ecosystem service values (ESV in 2007 US$ x 106 per year) estimated for each land cover category, and the overall change.

Landcover	Total ESV (2007 USD/ ha/year)*	Annual flow (2007 US$ x 10^6 per year)				% Change
		2000	2010	2020	2030	2000-2030
Lit	0	0	0	0	0	0
Cla	774	16.1	17.4	18.6	19.6	22
Sun	9,554	190.6	209.6	209.9	208.1	9
Smn	1,258	14.3	12.3	10.4	9.2	35
Riv	2,554	2.4	2.5	2.6	2.6	8
Aps	0	0	0	0	0	0
Urb	0	0	0	0	0	0
Agc	0	0	0	0	0	0
Lco	0	0	0	0	0	0
Total flow		223.5	241.9	241.7	239.7	7

The increase in the total ESV during the study period is presumably a result of the highest value assigned worldwide to the saltmarsh/ unconsolidated bottom cover. However, the analysis projects a significant decline (approximately 12%) in the ESV provided by the saltmarsh/forested mangrove class, representing potential high costs for the local inhabitants.

DISCUSSION AND CONCLUSIONS

The ecosystem service valuation issue is far from solved, mostly because there is not agreement on the ES concepts and classification, although the Millennium Ecosystem Assessment (MA) scheme is currently the most widely used despite its inconsistencies [26]. While the concepts evolve and criteria are unified, this classification system has been used for landscape management and valuation purposes, but more studies on mapping, modelling and spatial dynamics of the SE are required, to anticipate changes, understand processes and planning [1].

Depending on the extent or characteristics of the study area, in many cases, applications from remote sensing and GIS provide the only economically feasible way to gather regular land cover information to achieve the above mentioned goals [27,28]. Here, we use the potential information provided by multi-temporal Landsat TM data analyses to accurately map and analyze trends that can be used as inputs in the ES valuation, for further management processes regarding wetlands in southern Sinaloa.

Together with other analytical tools that allow some level of forecast, it is possible to produce indicators on the direction (from one class to another) and extent (ha) of LULC change in the future with Markov chains that output transition probabilities, a measure of the possible change. Here, this methodology forecast that littoral and saltmarsh wetland classes (both unconsolidated bottom and forested mangrove) are those with the highest probability of change in the near future. These results are consistent with findings from Robles et al. [24] for the northern coast of Nayarit, who found that saltmarsh, was the type of wetland most susceptible to change, particularly by effect of shrimp farm growing.

The total flow of ESV by wetland category was estimated based only on the land cover that resulted from the projection derived from Markov chains. The results indicate that the rates of change in the total ecosystem service value increase during 2000 to 2010 period, with the total flow of ESV increasing by 18 million (2007 USD), mostly as consequence of the increase in area of the unconsolidated saltmarsh cover. This positive trend in the total ESV flow contrasts with the trends found in other studies. For example, Kreuter et al. [29] and Zhao et al. [30] conducted assessments on ecosystem service values, and both studies found that the total annual ecosystem service value declined over the time and this reduction is attributed mainly to the effect of urban sprawl on land cover.

In this sense, our study suggests that even when there has been an urban growth during the study period, there is still no influence of this trend on the ESV provided by natural wetlands, and changes occur among wetland covers that supply different ES and their related values. In addition, it is important to note that the increase in unconsolidated saltmarsh is most likely a consequence of mangrove loss, but unlike other wetlands, saltmarsh is highly valued in other latitudes but not in Mexico, where this cover is considered unproductive, salty and bare soil. As a consequence, increases in this cover could be overvaluing the estimation. Future research to assess the ecosystem services value for this cover at a regional level is necessary to adjust the total flow. In fact, the use of estimates derived from Valdez et al. [24] was a good starting point to assess the values of ecosystem services in the study area, but also is required to conduct a primary study that focus on the evaluation of saltmarshes.

Based on the landscape trends observed from 2000 to 2010, a change in total flow is expected in the future, and in the projection for the year 2030, the ESV provided by the saltmarsh/unconsolidated bottom cover could decrease by approximately 200 ha, mostly as consequence of the increase in tourism infrastructure, which is projected to encompass approximately 40,000 hotel rooms and the services they require. It is also expected that the population will grow with urban expansion, leading to a decline in natural wetland covers and therefore, in the ecosystem services they provide. Focusing only on the ESV estimations for each wetland, the reduction of the saltmarsh/forested mangrove category over the time was the most important change in the study period, although changes affecting the value of mangrove coverage are due to interactions between natural wetlands instead of changes in land use. The provision of ecosystem services and the ESV of this wetland type could be altered if the negative trend continues. However, despite the relevance of this wetland cover, its contribution to the total flow is not significant (5% on average). Thus, its variation has a relatively small impact in monetary terms, as measured in the present approach.

The importance of ESV has been widely recognized as a useful tool to enhance land use planning [7]. Unfortunately, the underlying uncertainties and constraints in an ecosystem valuation using value transfer make it difficult to encourage policy makers to incorporate such valuation in environmental management studies, even when ESV estimation continues to be an efficient element to guide decisions for policy makers. Some of the limitations related to the use of value transfer in decision making could be addressed through greater interactions between researchers and policy analysts [31].

Although the described limitations exist, research using more accurate ESV methods that incorporate the spatial dimension, together with the best economic indicators at a regional level, are urgently needed to encourage policy makers to use ESV as an additional tool in their efforts to balance sustainable land use and regional development conditions [32].

ACKNOWLEDGEMENTS

We would like to thank Eva M. Tello and Rafael Hernandez for their valuable support.

REFERENCES

1. Millennium Ecosystem Assessment (2005) Ecosystem and human well-being: synthesis. Island Press, Washington, DC.
2. Troy A, Wilson MA (2006) Mapping ecosystem services: practical challenges and opportunities in linking GIS and value transfer. Ecol Econ 60: 435-449.
3. Wallace KJ (2007) Classification of ecosystem services: problems and solutions. Biol Cons 139: 235-246.
4. National Research Council (NRC) (2005) Valuing Ecosystem Services: Towards Better Environmental Decision-Making. National Academies Press, Washington, D.C.
5. De Groot R, Brander L, Van Der Ploeg S, Costanza R, Bernard F, et al. (2012) Global estimates of the value of ecosystems and their services in monetary units. Ecosystem Services 1: 50-61.
6. Halpern BS, Walbridge S, Selkoe KA, Kappel CV, Micheli F, et al. (2008) A global map of human impact on marine ecosystems. Science 319: 948-952.
7. Turner RK, Van den Bergh JCJM, Soderqvist T, Barendregt A, Straaten J, et al. (2000) Ecological economic analysis of wetlands: scientific integration for management and policy. Ecol Econ 35: 7-23.
8. Martinez ML, Intralawan A, Vazquez G, Perez-Maqueo O, Sutton P, et al. (2007) The coasts of our world: ecological, economic and social importance. Ecol Econ 63: 254-272.
9. Camacho-Valdez V, Ruiz-Luna A, Ghermandi A, Nunes PALD (2013) Valuation of ecosystem services provided by coastal wetlands in northwest Mexico. Ocean Coast Manag 78: 1-11.
10. Huber R, Rigling A, Bebi P, Brand PB, Briner S, et al. (2013) Sustainable land use in mountain regions under global change: synthesis across scales and disciplines. EcolSoc 18: 20.
11. Vitousek PM (1994) Beyond global warming: ecology and global change. Ecology 75: 1861-1876.
12. INEGI (2010) Anuarioestadístico de Sinaloa 2010: InstitutoNacional de EstadísticaGeografía e Informática: 553.
13. Berlanga CA, Ruiz LA, De la Lanza G (2010) Esquema de clasificación de los humedales en México. InvestigacionesGeográficas, BoletíndelInstituto de Geografía. UNAM 66: 25-46.
14. Pontius Jr RG, Millones M (2011) Death to kappa: birth of quantity disagreement and allocation disagreement for accuracy assessment. Int J Remote Sens 32: 4407-4429.
15. Fan F, Wang Y, Wang Z (2008) Temporal and spatial change detecting (1998-2003) and predicting of land use and land cover in core corridor of Pearl River Delta (China) by using TM and ETM+ images. Environ Monit Assess 137: 127-147.
16. Stewart WJ (1994) Introduction to the Numerical Solutions of Markov Chains.Princeton University Press.
17. Weng Q (2002) Land use change analysis in the Zhujiang Delta of China using satellite remote sensing, GIS and stochastic modelling. J Environ Manage 64: 273-284.
18. Eastman JR (2009) IDRISI 16: The Taiga Edition. Clark Labs, Worcester, Mass.
19. Barbier EB, Acreman MC, Knowler D (1997) Economic valuation of wetlands: a guide for policy makers and planners. Ramsar Convention Bureau, Gland, Switzerland.
20. Barbier EB (2007) Valuing ecosystem services as productive inputs. Econ Pol 22: 177-229
21. De Groot RS, Wilson MA, Boumans RMJ (2002) A typology for the classification, description and valuation of ecosystem functions, goods and services. Ecol Econ 41: 393-408.
22. Turner RK, Georgiou S, Fisher B (2008) Valuing Ecosystem Services: The Case of Multi-Functional Wetlands. London: Cromwell Press.
23. Ghermandi A, Van den Bergh JCJM, Brander LM, de Groot RS, Nunes PALD (2010) Values of natural and human-made wetlands: A meta-analysis. Water Resour Res: 46.
24. Camacho-Valdez V, Ruiz-Luna A, Ghermandi A, Berlanga-Robles CA, Nunes PA (2014) Effects of land use changes on the ecosystem service values of coastal wetlands. Environ Manage 54: 852-864.
25. Landis JR, Koch GG (1977) The measurement of

observer agreement for categorical data. Biometrics 33: 159-174.

26. Fisher B, Turner KR, Morling P (2009) Defining and classifying ecosystem services for decision making. Ecol Econ 68: 643-653.

27. Verstraete MM, Pinty B, Myneni RB (1996) Potential and limitations of information extraction on the terrestrial biosphere from satellite remote sensing. Remote Sens Environ 58: 201-214.

28. Seidl AF, Moraes AS (2000) Global valuation of ecosystem services: application to the Pantanal da Nhecolandia, Brazil. Ecol Econ 33: 1-6.

29. Kreuter UP, Harris HG, Matlock MD, Lacey RE (2001) Change in ecosystem service values in the San Antonio area, Texas.Ecol Econ 39: 333-346.

30. Zhao B, Kreuter U, Lia B, Ma ZJ, Chen JK, et al. (2004) An ecosystem service value assessment of land-use change on Chongming Island, China. Land Use Policy 21: 139-148.

31. Johnston RJ, Rosenberger RS (2010) Methods, trends and controversies in contemporary benefit transfer. J Econ Surv 24: 479-510.

32. Wu KY, Ye XY, Qi ZF, Zhang H (2013) Impacts of land use/land cover change and socioeconomic development on regional ecosystem services: The case of fast-growing Hangzhou metropolitan area, China. Cities 31: 276-284.

CHAPTER 17

COASTAL FLOODING IN THE SOLENT: AN INTEGRATED ANALYSIS OF DEFENCES AND INUNDATION

Matthew P. Wadey [1], Robert J. Nicholls[1] and Craig Hutton[2]

[1]Faculty of Engineering and the Environment, University of Southampton, Southampton SO17 1BJ, UK

[2]Geodata Institute, University of Southampton, Southampton SO17 1BJ, UK

ABSTRACT

This paper demonstrates a methodology for integrating existing models for the rapid simulation of coastal flood events across a large and varied case study area on the UK south coast. Following validation against observations from real coastal floods, synthetic events driven by realistic waves and water levels and the full range of failure mechanisms were modelled for a range of loadings to generate peak flood water depths and an overview of impacts across this spectrum of possible floods. Overtopping is relatively important compared to breaching as coastal floodplains are small. This modelling system supports multiple potential applications, such as planning flood warnings, coastal defence upgrade, and land use, including under sea-level rise. The concepts drawn from this study are transferable to similar coastal regions.

INTRODUCTION

Floods are an important long term risk to society. While society generally adapts to them, e.g., [1,2], this is usually accomplished in response to real events, which can include significant losses and even deaths. In Europe, developments in coastal flood risk management accelerated following the 1953 North Sea floods which killed 1836 people in the Netherlands, 307 in the UK and 17 in Belgium [3,4,5] and floods on the German Bight in 1962 when more than 300 people lost their lives [6,7]. In comparison with understanding at that time, coastal planners now benefit from a better understanding of the nature and degrees of exposure to flooding due to advances in coastal modelling e.g., [8]. Numerical simulation of coastal inundation is well demonstrated [9,10,11,12], although uncertainties, practical constraints, and the conceptual understanding of coastal flood systems can be limiting [13]; with poor flood incident management still resulting in widespread loss of life and other significant damages [14]. Hurricane Katrina demonstrated the catastrophic effects of defence failures in low-lying and populated coastal areas [15], while flooding in Bangkok has highlighted the impacts of flooding upon a large populous city, *cf.* [16]. The consequences of un-developed flood management measures were tragically apparent on the French Atlantic coast in February 2010 when more than 40 people died following failure and outflanking of coastal defences during the well forecast storm Xynthia [17]. In the hours before the event, warnings were available, but not

Citation: Matthew P. Wadey 1,*, Robert J. Nicholls 1 and Craig Hutton 2, Coastal Flooding in the Solent: An Integrated Analysis of Defences and Inundation, doi:10.3390/w4020430

specific to coastal flooding consequences, leaving authorities incapable of planning evacuation [18].

This study is differentiated from most flood risk research by focusing upon analysis of inundation over a large area where coastal floodplains are relatively small, and water flowing over defences (overflow and wave overtopping) may be more significant compared to breach (a reduction in the effective crest height). The latter failure mechanism is a particular threat for low-lying and expansive floodplains such as those on the east UK and Belgian, Dutch and German North Sea coasts and has received considerable attention [19]. Sea-level rise is increasing the probability of extreme events in most coastal regions e.g., [20,21,22], hence it is essential that methods are available to identify the likelihood and characteristics of flooding. In Europe, the EU Flood Directive (2007/60/EC) has determined that a framework is required to assess flood risk, including emphasis upon the frequency, magnitude and consequences of flooding. Few flood maps contain information on consequences [23], although flood risk assessment methodologies [24,25,26,27] are available to provide valuable insights to vulnerable defences and floodplain areas. Some countries benefit from national-scale assessments which make optimal use of the information available, e.g., [28], although places remain where flood dangers are unknown, or where spatial and temporal resolutions of broader assessments provide an abstraction of the reality of single coastal flood events.

High resolution datasets and numerical tools are currently available for many coastal areas, although are not routinely integrated; particularly where risks are not obvious or recently experienced This paper describes a method for such integration that can support coastal management measures such as forecasting and warning, defence planning, and land-use zonation. This is intended to provide outputs that are easily communicated to the public and flood managers. The Solent is selected as a case study; a shoreline on the south coast of England which comprises a large estuarine system and wave exposed open coast. Defended and undefended floodplains coexist, whilst flood consequences and risks are not well understood, and sea-level rise is increasing flood risk [20]. The aims of this paper are to:

1. Describe a coastal flood modelling methodology, where flood simulations account for real flood defence characteristics and a full range of defence failure mechanisms (ranging from simple overflow or outflanking, through to full breaching).
2. Simulate synthetic coastal floods within the Solent, to indicate flood event impacts under present and 21st century sea levels across this range of failure mechanisms.
3. Compare the flood consequences for different failure mechanisms in these floods.

A background to the Solent will now be described, followed by description of the methodology, results, discussion and conclusions.

THE SOLENT CASE-STUDY REGION AND INITIAL ASSESSMENT OF COASTAL FLOOD RISK

The Solent case-study is within the Environment Agency's Southeast Region. Alongside the South Downs, the Solent is a sub-region for which defences and flood warnings are managed; and contains two shoreline management plan (SMP) [29] areas (the mainland, and the Isle of Wight). SMP studies provide a framework for dealing with flooding and erosion by dividing sections of shoreline on the basis of coastal processes [30] and related defence systems. Figure 1 introduces the location of the Solent case study area, the areas broadly at risk of coastal flooding, and six 'sub-regions' for analysis described later in this paper.

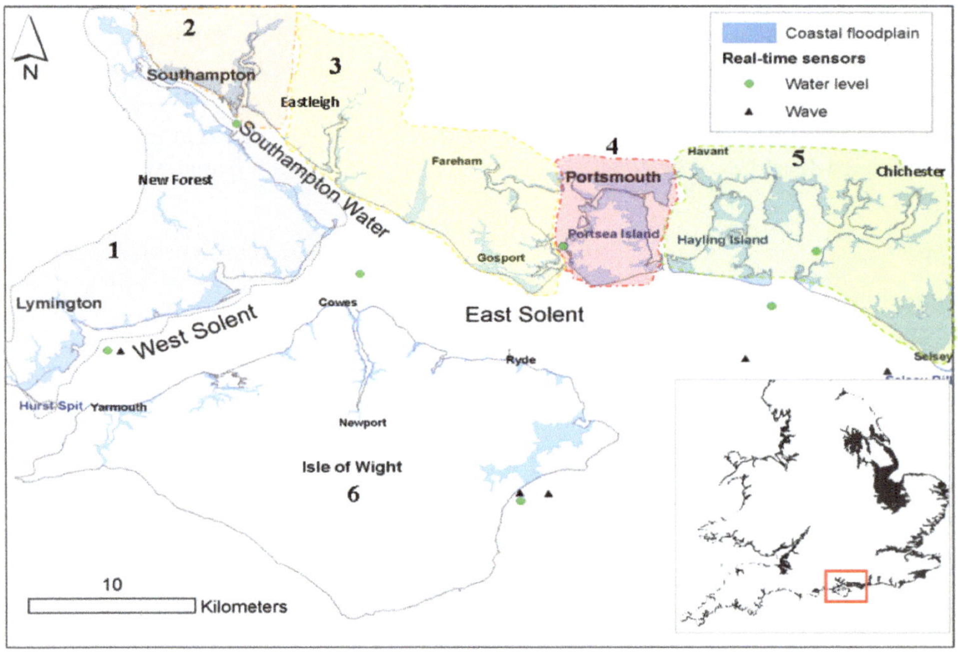

Figure 1. The Solent case-study, locations of wave and water level recorders, and the approximate 1 in 1000 year coastal floodplain (refer to Section 3.2). The base map (lower right) is an Environment Agency 1 in 1000 year indicative coastal floodplain for England and Wales [31]. Sub-regions used to describe flood modelling results later in the paper are: (**1**). The New Forest; (**2**). The city of Southampton; (**3**). The boroughs of Eastleigh, Fareham and Gosport; (**4**). The city of Portsmouth; (**5**). The boroughs of Havant and Chichester; (**6**). The Isle of Wight.

Much of the Solent is sheltered from south-westerly Atlantic waves by the Isle of Wight, and a managed shingle barrier at the western end known as Hurst Spit [32]. Storm surges in the Solent mainly occur as a result of low pressure systems that move from the Atlantic eastward over southern England [33] whilst smaller surges also occur as a result of large North Sea storm surge events transmitted into the English Channel through the Dover Strait [34]. Tidal residuals rarely exceed 1 m; with only a 0.33 m difference between a 1 in 10 and 1 in 1000 year water-level [35]. Coastal floods in the Solent during the 20th and early 21st centuries have been frequent but usually involving small water depths, and with no recorded loss of human life [36].

The form and extent of the estuary has largely been controlled by changes in sea level until progressive development and land reclamation began in the 18th century. This was most significant around Southampton Water and Portsmouth; the two cities having grown together to form an urban area containing over one million inhabitants. The tidal range increases from approximately 2 m at Hurst to 5 m at Selsey. The region is internationally renowned for its complex tides [37]. Double high waters occur and are particularly pronounced during large spring tides, and at mid-flood tide the tide is constant for about an hour (Figure 2). Meteorological induced sea level effects on the UK south coast are generally less severe than on the east and west coasts; although surge events [20,38] and Atlantic swells [39] have been associated with coastal flooding. A national assessment of flood risk [40] identified that the South coast would experience some of the largest increases in flood risk during the 21st century. It has been estimated that 24,138 properties in the Solent (excluding the Isle of Wight) are exposed to a 1 in 200 year coastal flood [41] with the largest concentration of risk in Portsmouth which contains more than 15,000 of these properties alone [41,42]. After London and Hull, Portsmouth may contain the greatest coastal flood risk for any city in the UK [43]. It is estimated that 25 percent of the city's coastal defences provide less than a 1 in 200 year level of protection (the indicative standard of protection for urban coastal areas in England and Wales) [44].

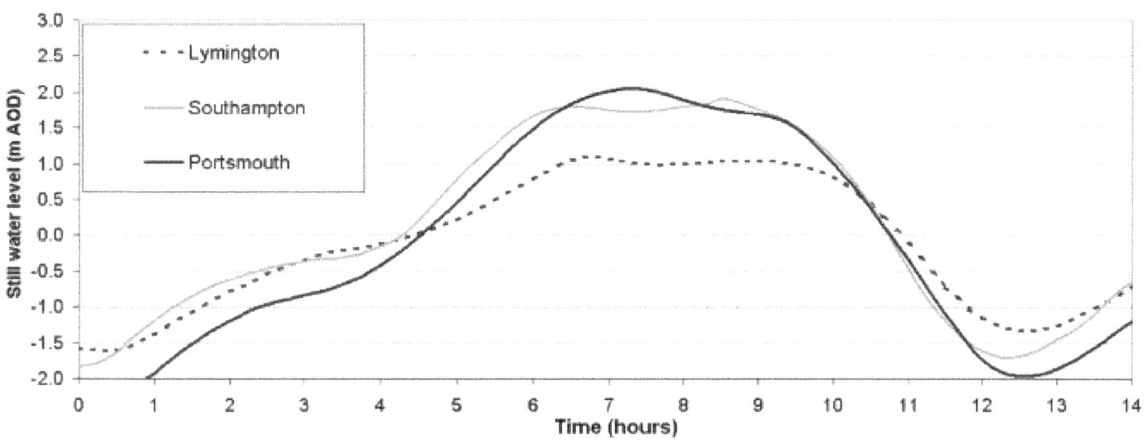

Figure 2. Water level time-series; recorded at three locations across the Solent (refer toFigure 1) during a typical spring tide.

Shown in Figure 2 are typical semi-diurnal lunar spring tidal cycles for three locations in the Solent. In Figure 3 are water levels recorded during a recent storm surge event. The Solent's defences are frequently subjected to hydraulic action, which due to the double high tides can be prolonged during storms. This effect is most prevalent in the central and eastern area of the region. The double high-water is significant at Southampton (which has almost no coastal flood defences). Portsmouth contains floodplains that are situated much lower (relative to extreme water levels) than Southampton, although there are flood walls and/ or beaches around the city. Defence design and floodplain characteristics are highly variable across the remainder of the region, as shown in this paper.

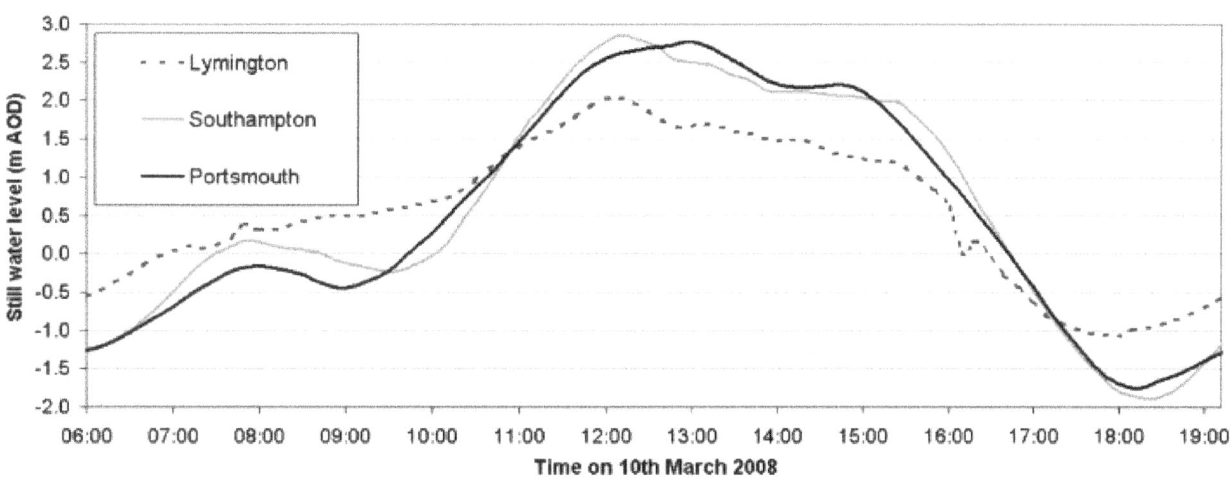

Figure 3. Water level time-series recorded during a storm surge event on the 10th March 2008.

METHODS

The modelling approach is summarised in Figure 4. Regional estimates of exposure to coastal flooding in the Solent were generated by combining planar water levels with a digital elevation model (DEM). This approximates coastal flood outlines and depths without considering defences or the dynamics of flooding. The second method simulates flooding more realistically by inclusion of hydraulic connectivity and mass conservancy; by coupling flood defence analysis and storm-tide water level time-series with a suitable numerical inundation model. Data layers were used to help prepare the model and view outputs, using geographical information systems (GIS). These included land-use maps, property datasets (containing commercial, residential, and unoccupied buildings), and aerial photography.

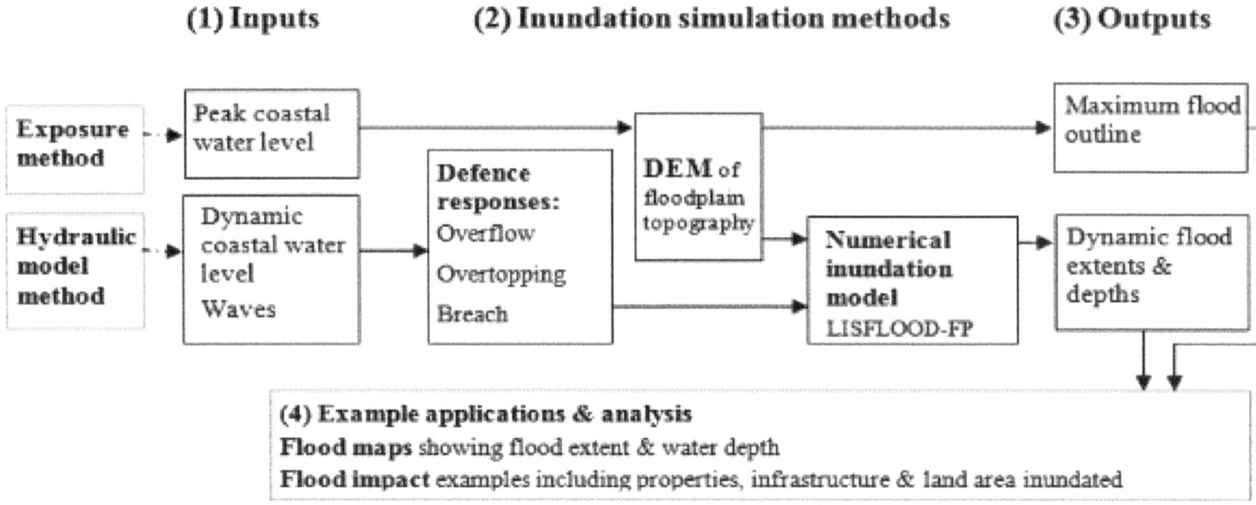

Figure 4. Method and model overview showing (**1**) inputs; (**2**) inundation simulation approaches; (**3**) outputs; and (**4**) example applications.

Regional Topography, Water Level and Defence Datasets

The DEM used for this work was mostly constructed from Light Detection and Ranging (LiDAR) surveys, processed in the format of a digital terrain model (DTM) (artefacts such as buildings removed). At the upper reaches of several tidal rivers the DTM was extended using a 10 m resolution topographic dataset derived from photogrammetry [45]. The LiDAR data was collected in 2007 and 2008 at 1 m and 2 m spatial resolution. Accompanying metadata indicates survey points are within ±0.15 m vertical accuracy [46], although an absolute error of ±0.30 m is often quoted [47,48]. For the hydraulic modelling method, shoreline flood defences were masked from the DEM, which was re-sampled from its original 1 m and 2 m resolution to 50 m cells; using a bilinear interpolation technique recommended for floodplain modelling [49,50]. This grid cell size allowed computational efficiency, and was commensurate with the available flood defence survey data (which was re-inserted onto the DEM). Large areas of the region comprise rural floodplain, and this resolution has been considered suitable in previous studies e.g., [51]; although benefits of finer floodplain grid cell sizes are acknowledged (Section 5.3).

Still water level return periods in the Solent have been periodically updated, and older methods for estimation of extreme still water level probabilities [52,53,54] are noted to significantly over-estimate return levels on the UK south coast [55]. For example, the 1 in 200 year water level definition shows variations of more than 0.1 m between those published in 2007 [56] and recent analysis of tide data at Southampton and Portsmouth [35]; one reason that quantification of property exposed to the 1 in 200 year flood in this work will vary in comparison to previous estimates mentioned in Section 2 of this paper. The latest available still water level return periods are from a national study which applied the Skew Surge Joint Probability Method around the UK coast [57] (e.g., Table 1). These latest levels were used in this work, and prescribed spatially to simulations of flood events by methods used for the Environment Agency's flood zone mapping and a previous flood risk assessment in the Solent [44] (see Figure 5). In relation to Portsmouth this equates to extreme water levels that are approximately 0.8 m less at the western end of the region (Hurst) and to 0.5 more at the eastern extent (Selsey Bill).

Table 1. Return still water levels at Portsmouth in 2008 [57], mean high water springs, and the still water level recorded during a recent storm surge. The water level datum referred to (mAOD) is approximately mean sea level.

Return period (years)	1	10	20	50	100	200	1000	MHWS	10th Match 2008
Elevation (mAOD)	2.56	2.81	2.88	2.98	3.05	3.12	3.28	1.97	2.77

Regional sea-level rise in the past decade has been approximately 1.7 mm per year [58] hence the most recently derived return periods do not require significant adjustment.

Wave data was compiled from different sources, to enable wave loading scenarios to be combined with regional sea level extremes. Significant wave height return periods were available at measured sites (Figure 1) [59]; and fetch-limited conditions were calculated in all areas [60]. Records of the largest swell wave heights and periods that have occurred since records began

[46] were also noted for exposed coastline east of Portsea Island. Defence datasets describing condition and structural elements (such as roughness, berms, and crest heights) were available across the region. Information was also gained from the DEM, aerial photography, photos of defences, literature and site visits. Crest heights were mostly from real-time kinematic (RTK) GPS surveys with a vertical accuracy of ±0.03 m [46]. Recent condition grade scores (between 1 for excellent, and 5 for poor) were supplied by local authorities, allocated by coastal engineers to defences in England and Wales during routine inspections, and which can be mapped to defence performance under loading [61,62].

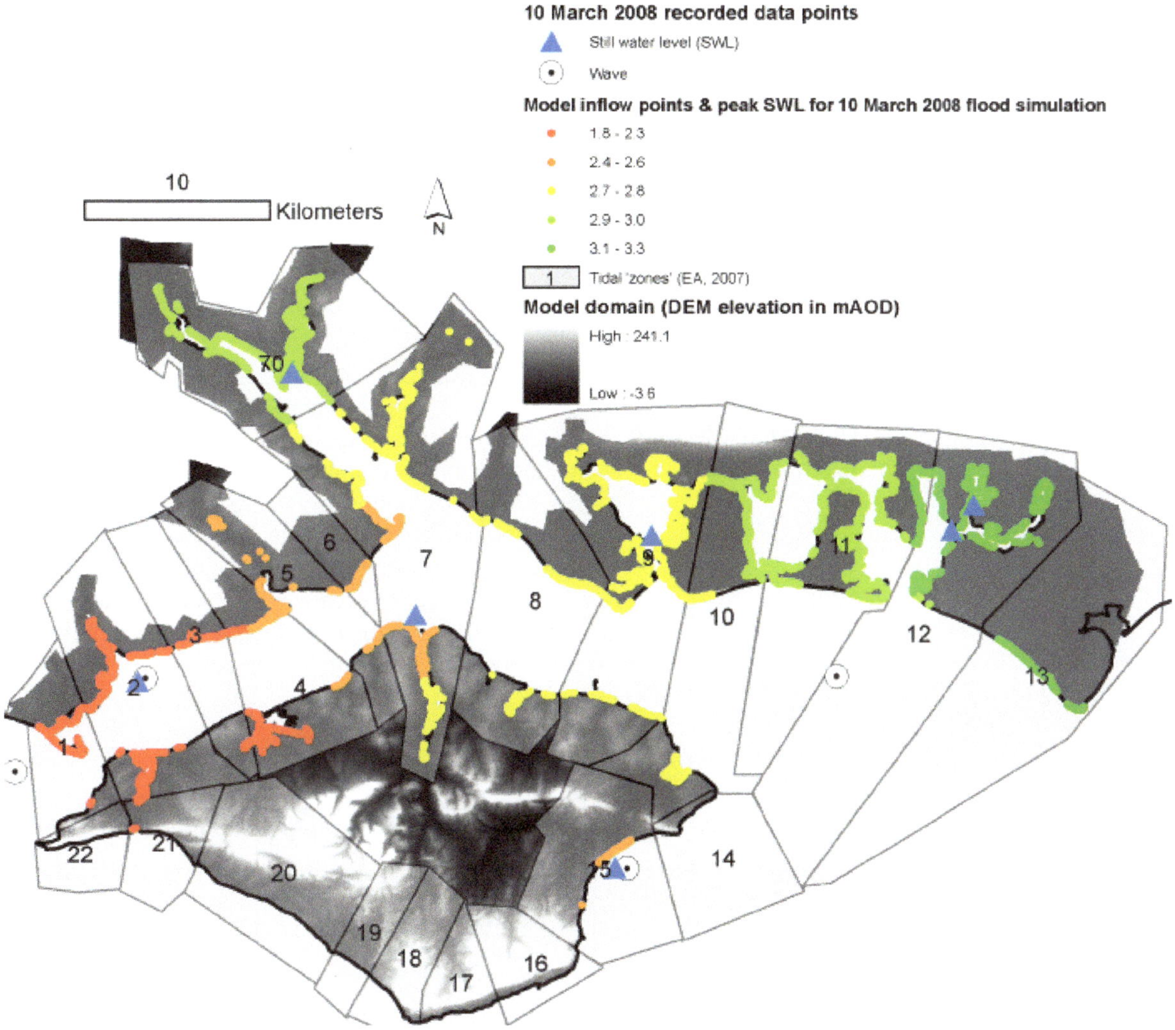

Figure 5. The computational domain and boundary still water levels for the 10th March 2008 regional flood simulation. For this example water levels at inflow points were spatially (linearly) interpolated between data recorders. For simulations of synthetic flood events (Section 4) the zones shown were used to allocate shared still water level characteristics to the model boundary (refer to Section 3.1). Water level time-series for the hydraulic modelling were based upon a Lymington storm-tide curve for zones **1**–**6**; Southampton Water for zones 7 and 70; and Portsmouth for zones7–**15** (refer to Figure 3).

The Planar Water Level Method

Regionally uniform synthetic coastal events were modelled across a range covering normal tidal levels to extremes which included mid-range estimates of 21st century sea-level rise [63]. The planar water level method defined maximum flooding for given levels of still water (and allowance for 2 m wave run-up to define properties exposed to flooding on the open coast of Hayling Island). This identified 99 hydraulically discrete and independent flood 'compartments' covering 120 km^2 of land. These compartments were separate on the basis of a 1 in 1000 year regional planar water level plus 0.3 m allowance for vertical data errors.

Hydraulic Modelling

Where inundation from defence failures was simulated, peak flow was calculated at the boundary of the model. This was spread to adjacent cells by solving of a continuity and momentum equation in each direction using the inertial version of the raster-based inundation model LISFLOOD-FP [64]. LISFLOOD-FP originated as a simple storage cell model solving an analytical approximation of the 2D shallow water equations [65,66]. Use of high spatial resolution and good quality topographic data has allowed similar results to full 2D codes (for sub-critical gradually varied flows only), and LISFLOOD-FP has been previously been used for modelling coastal floods [9,12,51,67]. At higher grid resolutions the requirement for small time-steps to allow stable and realistic simulations of flow [66,68] prompted incorporation of an inertial formulation of the two dimensional shallow water equations into LISFLOOD-FP, which has been noted to significantly increase computational efficiency [64,69]. Flow between neighbouring floodplain cells is calculated as in Equation (1):

$$q^{t+\Delta t} = \frac{q^{t} - gh^{t}_{flow}\Delta t \dfrac{\Delta(h^{t}+z)}{\Delta x}}{\left(1+gh^{t}_{flow}\Delta t n^{2}\left|q^{t}\right| / (h^{t}_{flow})^{10/3}\right)} \quad (1)$$

where q is the flow per unit width, h_{flow} is the depth of water available to flow, Δt is the model time step, Δx is the model grid resolution, z is the bed elevation, h is the water depth, and n is the Manning's friction coefficient. Water depths are updated using Equation (2):

$$h^{t+\Delta t}_{i,j} = h^{t}_{ij} + \Delta t \frac{Q^{t}_{xi-1j} - Q^{t}_{xij} + Q^{t}_{yij-1} - Q^{t}_{yij}}{\Delta x^{2}} \quad (2)$$

where Q is the discharge, h_{ij} is the depth of water in the cell (i,j). The stability criterion for this numerical model is given by the Courant–Freidrichs–Levy condition for shallow water flows such that the stable model time step, Δt, is a function of the grid resolution, Δx, and the maximum water depth, h, within the domain:

$$\Delta t_{max} = \alpha \frac{\Delta x}{\sqrt{gh_{t}}} \quad (3)$$

where α varies between 0.2 and 0.7. The reader is referred to Bates *et al.* (2010) [64] for a fuller explanation of this model.

Importantly, when coupled to the defence response models this model is rapid enough for real-time regional simulations up to the peak water level, driven by applying water level time-series at the model boundary. Almost 5000 potential inflow locations were defined along 246 km of intertidal shoreline comprising point vectors with data attached (including defence crest height floodplain height at that location, and parameters for overtopping and breach analysis). These were placed in cells on the seaward edge of the 50 m resolution DEM, covering all flood compartments. A line of points was also placed in cells immediately inland of the defences for wave overtopping failures to be simulated. Water level time series at these locations were based upon tide-surge water level time-series (see Figure 3), adapted according to defence responses if necessary. Failure of defence systems is quite rare, although outflanking, overflow, overtopping and breach were considered.

Scope for calibration by varying surface friction values was limited by a lack of accurate flood observations, although boundary conditions (defence failure, waves, water level, *etc.*) dominated flood inundation modelling results. A uniform representative composite value of n = 0.035 was used, based on the range of surface roughness values found across the region. Some allowance was made for DEM error by excluding analysis of flooding in cells where water depths were less than 0.05 m [70]. As shown in Table 2, flood simulations over a relatively large area could be run quickly, even using a standard PC.

Table 2. Summary of the model scale and run-time (for a standard 2.5 GHz desktop PC).

Model	Domain size		DEM resolution (m) and file size (MB)		Number of inflow points	Time-step (seconds)	Run-time to simulate 14 hours of real-time flooding (minutes)
	Km²	columns, rows					
Solent	105.7	1227,846	50	3.87	4909	4.2–6.3	15
City of Portsmouth	15.2	150, 168	50	0.01	595	5.4–8	<1

Overflow

Overflow failure is when the still water level exceeds the crest height of a defence (or of the land if there is no defence). For flood simulations comprising no waves an estimate of overflow during a flood event was calculated by applying still the storm-tide water level time-series to inflow points lying seaward of defences in the DEM. When wave conditions were included, the original peak still water level was routed through boundary cells, although with water level time-series adjusted for the duration that the still water level could theoretically exceed the defence crest height. The latter method is most relevant to shorelines where defence crests are at a similar elevation to the ground level of the surrounding floodplain; or where seepage, gaps, and alongshore variations in crest heights (for any 50 m defence section) may allow a greater inflow.

Overtopping

Flooding due to overtopping is when the combined effects of water levels and waves results in water entering a floodplain at a faster rate than it can drain away. This mode of failure can occur when the still water level is below the height of the defence crest or the floodplain. To generate boundary conditions for the inundation model, the mean overtopping discharge, percentage of overtopping waves, and run-up were calculated at stages of the tidally varying freeboard using empirical methods from the EurOtop Manual [71]. For sloping structures the formula used was that by Owen (1980) [72]; TAW (2002) [73] for beaches; and for vertical and composite sea walls more iterative formulae (to discriminate between pulsating, impulsive and broken wave conditions, according to the standard EurOtop guidance). A bathymetry model was constructed using nearshore sounding data [46] to approximate the regionally variable effects of wave shoaling upon significant wave heights. The duration of overtopping events was delineated by time intervals where mean overtopping rates or run-up exceeded the known capacity of defences. If for any inflow point an overtopping event was simulated; the onset and termination of floodplain discharge was defined by exceedance of the run-up two percent value (at sloping defences) or a mean overtopping volume discharge of greater than 20 l/s/m. When defence crest freeboard was negative and the floodplains below still water level; the still water level time series was simply routed through the failed boundary cell (although restricted to the calculated overtopping duration). For most floodplains the peak (or equilibrium) water level in the overtopped boundary cell was limited to the seaward still water level. This limit was higher for floodplains above still water levels, based upon prior analysis of run-up and overtopping volumes in the relevant flood compartments. The results in this paper assumed a peak significant wave height (H_s) and mean period (T_m) to occur for the duration of an overtopping event; although this could be adjusted to account for more dynamic conditions if required. At each inflow location this was converted to a vertical change in water level expected to occur in the lee of defences during a single tidal cycle; and added to the DEM floodplain level in the boundary cell, forming the peak level in a modified time series of water levels.

Analysis of overflow and overtopping was applied to over 224 km of the region's shoreline, including beaches, embankments and vertical sea walls. It was found that approximately 116 km of these defences rise far enough above the floodplain (whilst preventing seawater from reaching hinterland) for breaching to be relevant.

Breach Analysis

Breaching, although rare, is considered because this type of defence failure is recognised to greatly increase the consequences of a flood event, especially for large flood plains [19]. Analysis was allocated to defences where breaching is relevant, allowing for the identification of possible regional 'weak spots' and information for the modeller to route water through boundary cells if chosen. A simplified reliability analysis generated fragility curves for three selected failure modes. This was facilitated by software developed under FLOOD*site* Task 7 [74], which enables a Level III reliability approach (*i.e.*, Monte-Carlo simulation), (see [27] for a more thorough description). Three breach failure mechanisms were selected from a much wider inventory [75]. These were: (1) analysis of damage initiated by overflow and/or overtopping [76]; (2) failure caused by water internally eroding defences (piping) [77]; and (3) critical erosion of shingle beaches in response to wave and water level loadings [78]. The former two failure mechanisms were analysed using condition grade scores and structural descriptors to imply defence strength and stability [79]. Topographic profile surveys were used for analysis of shingle beaches (which are important natural and managed defences in many areas of the Solent). Methods exist to calculate progression of breaching, and subsequent release of water onto the floodplain but any deterministic prediction remains highly uncertain [80,81]. Hence the modelling results in this paper take a simplified approach with breaches assumed to be broken down to floodplain level and water flowing through for the entire tidal cycle. Section 3.4.2 (this paper) describes inundation simulations where specific breach locations were approximated for a validation case study; whereas results inSection 4 describe flood simulations where breaching scenarios consider all shoreline defences to have been removed (and wave height reduction coefficients removed where foreshore usually alleviates overtopping). This allowed a simplified comparison with overtopping scenarios, and a more realistic worst-case scenario than the planar-water level results. There is some discussion of breach probabilities during an extreme flood event simulation in Section 4.2.

Model Validation

A model is said to be validated if its accuracy and predictive capability in the validation period have been proven to lie within acceptable limits or errors for a particular purpose [68]. However, for many coastal areas information to validate inundation models is not available. Flood simulations were compared with observations gained from two actual floods: (1) a regional event on 10th March 2008, which comprised widespread although mostly minor flooding across the Solent; and (2) an event in December 1989 where inundation resulted from overtopping and sea wall breaching between Hurst Spit and Lymington.

10th March 2008 Regional Flood Event

On the 10th March 2008 an approximately 1 m storm surge and large waves resulted in widespread flooding across the Solent and other coastal areas in the English Channel [82]. The region's cities escaped major inundation, although an example of a near miss in Southampton is shown in Figure 6. Observed flood extents were in the range 0.01–0.03 km^2 with peak flood water depths mostly less than 1 m. An exception was more severe flooding in eastern extent of the region at Selsey (located at the largest and most easterly floodplain shown in Figure 1), where the only known breach of defences occurred in the region that day due to waves flattening an 800 m section of shingle barrier beach [83]. At the time, the peak still water levels were approximated as a 1 in 20 year extreme at Portsmouth and a 1 in 50 year event at Southampton [20]. Waves and water level observations from the event were available from real-time data recorders used for regional coastal monitoring (Figure 1) and information collated by the Environment Agency [84]. These were spatially interpolated between recording stations and applied to the model boundaries (Figure 5). Outputs were compared with observed flooding in GIS. Observations were interpreted from photos, descriptions and reports, e.g., [84] and close examination of a high resolution version of the floodplain DEM which enabled flood extents and depths to be (approximately) verified and reconstructed.

Previous coastal flood modelling studies [9,11] have compared modelled and observed coastal flood extents using a fit measure (F_A) which comprises the intersection and union of predicted (E_p) and observed (E_0) flooded pixels. A value of zero corresponds to no agreement, a value of 1 to perfect agreement:

$$F_A = \frac{E_P \cap E_0}{E_P \cup E_0} \tag{4}$$

Figure 6. High water at Southampton docks on 10th March 2008 [85].

Table 3. Summary of the 10th March 2008 model validation.

Test type	Criteria	Result
Detection of observed floods	The model indicated flooding at 29 out of 30 locations where floods were reported (the location missed was on a tidal river where additional sources of floodwater may have occurred). At 27 locations, some description of flooding was available to allow comparison with the model; which is summarised below.	
Flood extent fit scores	Mean F_A value (for the 27 compared locations) over-predictionsunder-predictionsneither definitive over or under-prediction	0.656813
	Number of locations with a good fit (F_A > 0.75) 2 were slight over-predictions, 1 was a slight under-prediction Number of locations with a moderate fit (F_A = 0.50 to 0.75) 2 over-predictions, 3 under-predictions Number of locations with a poor fit (F_A < 0. 50) 2 were over-predictions, 4 were under-predictions	**12** 7 **8**
Flood depth	All modelled depths were within 30 cm of observed flood depths.	

For each measurement of F_A it was noted whether the model under or over-predicted flooding. Over-prediction or under-prediction by the model was considered negligible unless more than one full flooded cell lay outside of the observed extent. Hence in the case of a good model fit (F_A > 0.75) not all values qualified as over or under-predictions. Limitations to this exercise are acknowledged; such as the use of simplified boundary inputs, the relatively large pixel size (50 m) of the model, uncertainty in the timing of the observed flood data, and interference from other flood sources (during the real event). Flood depth was also noted, using photos and the DEM. The uncertainties from this method meant that if the modelled flood depth was within 30 cm of the observed depth (at any location), this was regarded as acceptable. The results of this exercise are summarised in Table 3.

The mean F_A value was reduced considerably by over-prediction of flood extent at two small urban floodplains, where the F_A values were 0.10 and 0.32. The former of these values was at a location where waves overtopped a shingle beach at the southeast corner of Hayling Island, the latter was overtopping of sea walls at Old Portsmouth (on the southwest corner of Portsea Island). These over-predicted floods comprised water depths of less than 20 cm. It is possible that these were reasonable predictions of overtopped volumes although in reality ponding of overtopped water may be lessened by floodplain drainage and the presence of surface features such as buildings (neither of which were included in the model). Reducing wave shoaling coefficients in the overtopping model (from approximately 0.75 to 0.60) improved these predictions; although the original values were retained for further modelling on the assumption that drainage may be limited in extreme events whilst shoaling may be expected to change with foreshore and nearshore sediment dynamics during storms. Furthermore, these variables are likely to be negligible in comparison to significant defence failures.

Specific locations of the properties affected by this event were not available, although a post-event report [84] suggested that 30 properties in the region were flooded. Viewing photographs taken at 24 of the flooded locations suggested that at least 120 buildings could actually have been in contact with flood waters (see Figure 7). There is uncertainty over how thoroughly floods were reported, whilst these numbers exclude flooding and evacuation of a holiday park at Selsey [86]. The modelled flood outline of this event contained more than 2000 properties, more than half of these accounted for within the shallow flooded cells in the aforementioned over-predicted flood extent on southeast Hayling Island. Filtering these results by the number of properties in grid cells with flood water deeper than 1 m decreased this estimate of inundated properties to 162; and to only 38 with water deeper than 1.5 m. It is probable that flood waters threatened many more properties than reported, but walls, fences, raised floor levels, and flood prevention measures such as sandbagging prevented significant flooding to interiors of buildings.

Figure 7. Flood waters surround properties in Emsworth, Borough of Havant near high water on 10th March 2008, [87].

17th December 1989 Flood Event (Breach Failure)

A historic flood was used to evaluate the model's ability to simulate larger floods resulting from significant defence breaching. Such an event occurred on the 17th December 1989 at a mostly rural flood compartment between Hurst Spit and the town of Lymington, on the west Solent. Rapid onset and widespread coastal flooding resulted in environmental damage and severe inundation of 10 properties in the rural area of the floodplain; with overtopping also affecting 50 properties in Lymington [88]. Measurements of wind speed, peak water level, breach locations, and defence heights are available for the event [88] and were used to model flooding over a single tidal cycle. Variations of boundary inputs were used to run several flood simulations, and these were compared to observed flood extents producing F_A values between 0.7 and 0.8. However, these F_A values do not take into account the two notable discrepancies between the present day DEM and the reported flooding of December 1989 (as noted in Figure 8). The landfill area is dynamic and flooding cannot be assessed, while in the vicinity of Milford on Sea, the breaching of Hurst Spit may have sealed a channel, protecting Milford-on-Sea from flooding. It is possible that other areas of the floodplain surface may also have altered between the time of the event and collection of the recent LiDAR data; although the old defence data (which is less detailed than the present-day dataset), nature of breaching, and wave data provides greater uncertainty to the model inputs for this validation exercise. As with the 10th March 2008 validation case-study, filtering properties within grid cells of certain depths provided a closer comparison with reported flooding. Counting only properties where modelled flood water depths exceeded 0.5 m indicated 104 properties to be inundated; depths of greater than 1 m indicated 47 properties to be flooded (60 properties were reported to have actually experienced flood damage [88]).

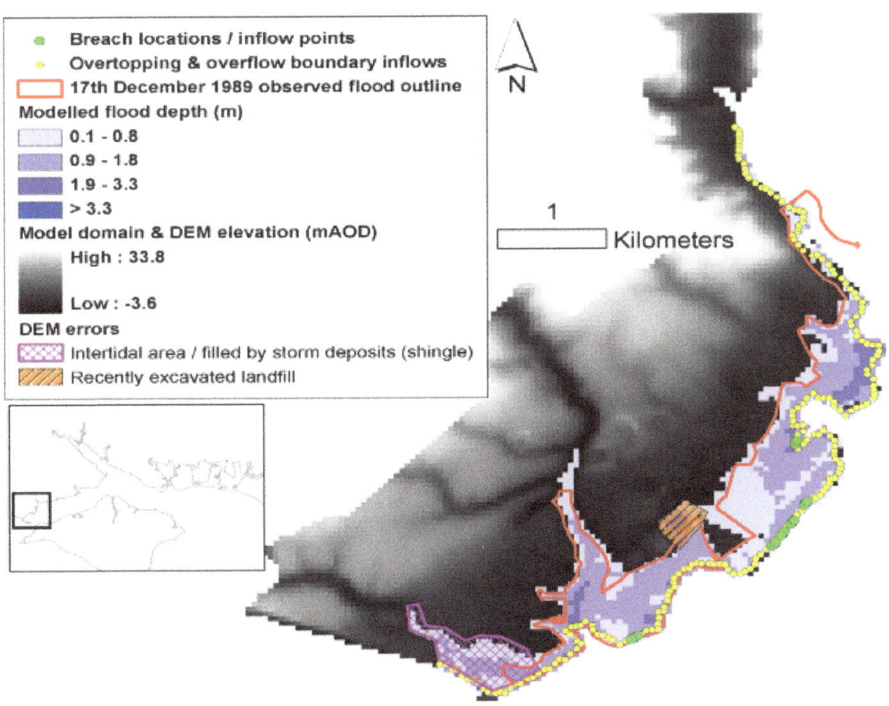

Figure 8. The 17th December 1989 flood simulation and observed flood outline. A boundary water level of 2.1 mAOD recorded at Lymington [88] was applied across the entire site for this simulation. This may be a relevant approximation since the breach of Hurst Spit that day

e.g., [89] would be likely to have altered local hydrodynamics [90,91] (whereas present-day event flood simulations assume Hurst Spit stays in place, and that tidal range/extreme water levels increase eastwards across the site). Over-washed material from the storm may also have blocked the intertidal area shown (at Milford on Sea) complicating comparison between the observed and modelled floods.

The fragility curve methodology and recent defence data was partly verified by this event. Information that indicated the condition and height of the sea wall in 1989 [88] produced breach failure probabilities ranging between 0.8–1.00 when subjected to the 17th December 1989 loadings (*i.e.*, the reliability analysis indicating definitive exceedance of the limit state for several sections). The same loadings applied to the present-day sea wall data indicated both minimal inundation from overtopping, and no definitive defence section failures, although breach probabilities were between 0.6 and 0.9 for some sections. This site did not flood on the 10th March 2008, even though the water levels exceeded those of the 17th December 1989.

RESULTS

Performance measures verified that the main components of the model (DEM, defence data, *etc*) are sufficiently accurate for modelling hypothetical coastal flood events. This section firstly focuses upon simulation of a 1 in 200 year regional coastal still water level flood (indicative of the standard of protection for urban coastal defences in England and Wales and definition of Environment Agency flood zone maps). Secondly, inundation across a range of coastal flood simulations is described. This indicates model sensitivity and uncertainty related to boundary water levels, and provides a regional overview of potential flood event impacts at present and higher sea levels. In this paper, the inundation simulations driven by defence failures are in a deterministic form, rather than comprising a sampling-based flood system analysis e.g., [26]. Inundation scenarios modelled for each still water level include: (a) the planar water level flood outline; (b) a worst-case hydraulically modelled flood that includes breach of all defences; and (c) three categories of wave loading simulated with the hydraulic model although *without* breaching (these include no waves, an annual extreme, and the maximum possible wave conditions) hence accounting for overflow, overtopping and outflanking failures only (for one tidal cycle). The still water levels (and definition of flood event extremity) applied to the model boundaries were based upon the extreme water level analysis described in section 3.1 and Figure 5. The water level time-series applied to the inflow points simply comprised shifting storm-tide curves (Figure 3) to the relevant peak still water level.

Simulation of an Extreme Present-Day Flood Event

Table 4 summarises the results of a 1 in 200 year water level flood. As expected e.g., [9], hydraulic modelling (even with extreme wave overtopping and breach scenarios) estimated less flooding than calculated by the planar water level method. The flood simulation using an annual wave extreme suggests that more than 10,000 properties could be impacted by this water level event. The model validation case studies indicated that waters exceeding 1 m depth could be a more useful indicator of the number of buildings that experience flooded interiors. This criterion suggests that for the synthetically modelled floods, increasing wave loadings from the annual to maximum extreme (and with no breaching) almost doubles the number of properties that would be significantly inundated. Inclusion of breaching (of all the region's defences) in the hydraulic flood simulation method almost trebles the effects from waves alone (for the count of properties inundated to greater than 1 m depth).

Table 4. Modelling results for a 1 in 200 year water level coastal flood in the Solent (refer to Figure 1 for locations).

REGION (seeFigure 1)	Location	Exposure (numbers in brackets is land area inundated to >1m depth)		Properties flooded by hydraulic simulations Total (number in brackets are properties inundated to >1 m depth)			
		Floodplain area (km²)	Properties	Breach of all defences and max. waves	Max. waves	1 in 1 yr waves	No waves
1	New Forest	14.1 (6.6)	736 (134)	639 (147)	591 (112)	430 (55)	329 (34)
2	City of Southampton	2.0 (0.1)	2,148 (249)	978 (21)	973 (21)	919 (21)	715 (10)
3	Eastleigh, Fareham & Gosport	7.2 (2.4)	1,789 (106)	1,395 (109)	1,321 (93)	1,163 (91)	564 (52)
4	City of Portsmouth	13.8 (4.9)	14,055 (3,483)	10,922 (2,734)	6,437 (459)	4,660 (110)	1,923 (8)
5	Havant & Chichester	29.8 (13.2)	3,781 (1,657)	3,168 (613)	2,995 (448)	2,602 (329)	1,033 (90)

6	Isle of Wight (all)	11.4 (5.6)	617 (155)	505 (114)	377 (67)	354 (67)	255 (25)
Total	**Solent**	78.3 (32.8)	23,126 (5,784)	17,607 (3,728)	12,694 (1,200)	10,128 (673)	4,816 (219)

In the city of Portsmouth, a substantial amount of the region's property is threatened by flooding under all loading scenarios. However the city's flood defences greatly reduce the number of properties exposed to more than 1 m flood depth; and the extreme wave and overtopping and breaching scenarios are required to significantly increase the count of buildings likely to be severely flooded. The reliability analysis indicated that loads associated with the 1in 200 water level (combined with waves) would generate low probabilities of breach for most of the city's defence structures. The probability of failure from the fragility curves was less than 20 percent for defences protecting two of the largest clusters of property in the central south and mainland areas, although failure probabilities are higher for sea walls on the east coast of Portsea Island. In the city of Southampton the added effect of waves with still water level expands the flooded area significantly, despite the fetch-limited nature of the shoreline. However, only 21 properties in the city are inundated to greater than 1 m depth. Hence, at present sea level Southampton is likely to be subject to disruptive impacts from coastal flooding, rather than significant risk to life and property.

Note that 531 properties were flooded to greater than 2 m depth in the case of an extreme breaching scenario (and approximately 130 for maximum and annual wave scenarios). The majority of these are in Portsmouth (311 properties) under the breach scenario; whilst most are in Chichester and Havant (102 properties) when applying only the maximum waves scenario (to generate extreme overtopping-dominated defence failures). There are also substantial areas of floodplain where flood depths are between 1 m and 2 m, which can be particularly dangerous when floodwaters are fast moving [92]. The extraction of this information from the model indicates pockets of particularly high damage potential and risk to life. Inclusion of widespread breaching is shown to increase regional flooding impacts substantially (Table 4), although only a relatively small length of coastline actually appears at high risk of this kind of failure. Locations with the highest probability of breaching are mostly in rural areas on tidal rivers and in the harbours. Simulating inundation through defence sections for which the failure probability exceeded 90 percent (along approximately 35 km of defence line) is not shown here, although generated a similar inundation extent to that of the extreme wave overtopping scenario. Slightly greater floodplain water depths resulted from inclusion of these breaches, although many of these defences are likely to have already significantly overtopped prior to breaching. Such failures would inevitably increase the consequences of a flood event, although the overall impacts would be dependent upon factors such the timing of breach initiation and growth, and repair measures.

Coastal Flood Simulations Across a Range of Loadings

Flood simulation results across a range of still water level are shown in Figure 9a, Figure 10a and Figure 11a (all flood simulations assumed the present defences). Also shown for each measure of inundation, is the proportion of flooding caused by non-breach defence failures in comparison to inundation simulations where all defences were removed (Figure 9b, Figure 10b and Figure 11b). These results indicate that substantial land areas and numbers of property are threatened with flooding from frequently occurring water levels, with defences playing a crucial role in the magnitude of these consequences. The number of properties in the modelled flood outlines increases steadily with increasing boundary water levels. The difference in the total number of flooded properties and land area when comparing the hydraulically simulated floods and the planar water level (exposure) floodplain indicates that much of the wider floodplain may not be hydraulically well-connected over a single tidal cycle. At the same time, the effects of overtopping can be significant and these results show that such floods can almost match the planar water level estimates for properties flooded to greater than 1 m depth (Figure 10a). These results show that for some individual flood compartments the number of properties inundated by the non-breach defence failure mechanisms increases most distinctly within the (approximately) 10 cm band of water levels either side of the 1 in 200 year extreme (because the protective capacity of many coastal defences is exceeded in this range of loadings). Using the 1 m depth criteria shows that for the Solent as a whole, there a particularly large increase in the count of flooded properties beyond the 1 in 200 year still water level.

Figure 9b and Figure 11b show a general increase (in response to increasing still water levels) in the percentage of inundation that arises from overflow/overtopping failures in relation to breach. This is expected as defences become increasingly overwhelmed beyond their protective capacity due to rising still water level boundary conditions allowing larger and prolonged discharges during flood events. However, for the 1 m depth criteria (Figure 10b) the potential for breach to exert the greatest flood damages across the Solent is most apparent for small water levels (where breach is unlikely and significant overtopping not possible), and a wide range of present-day extremes. Beyond present-day still water level extremes, the gap between breaching and overtopping impacts is greatly diminished.

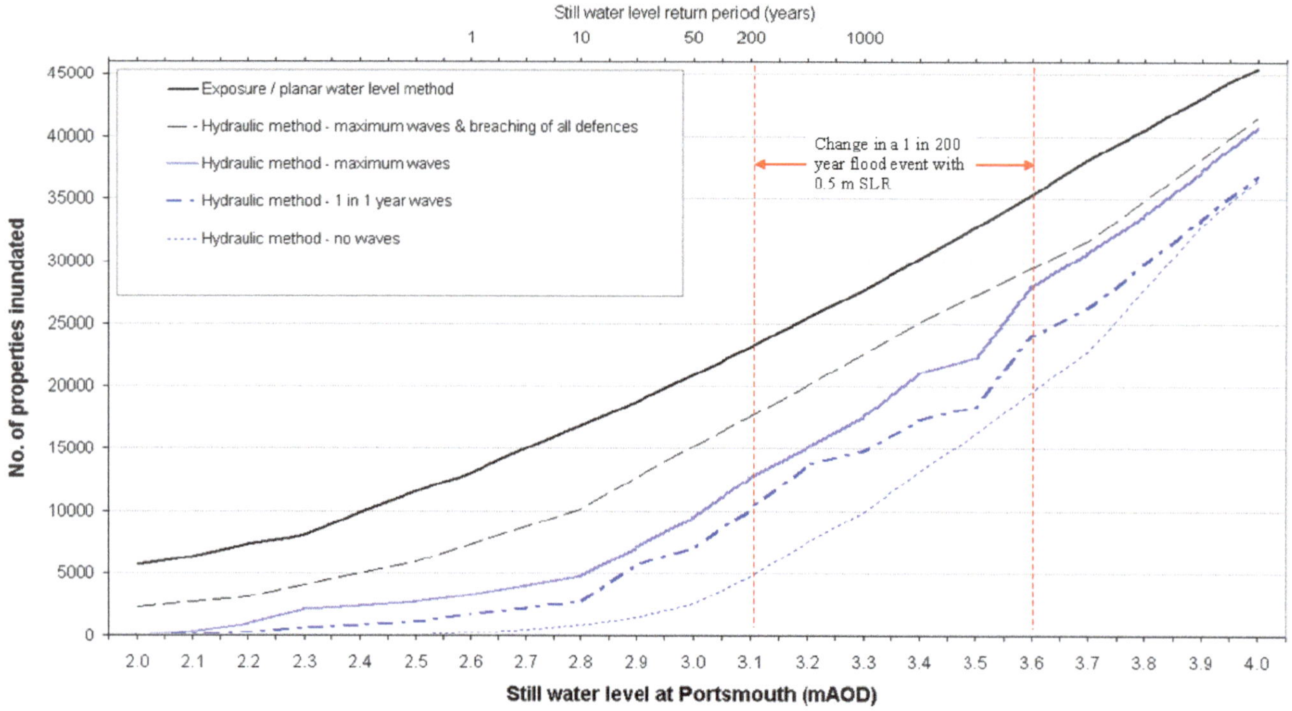

(a)

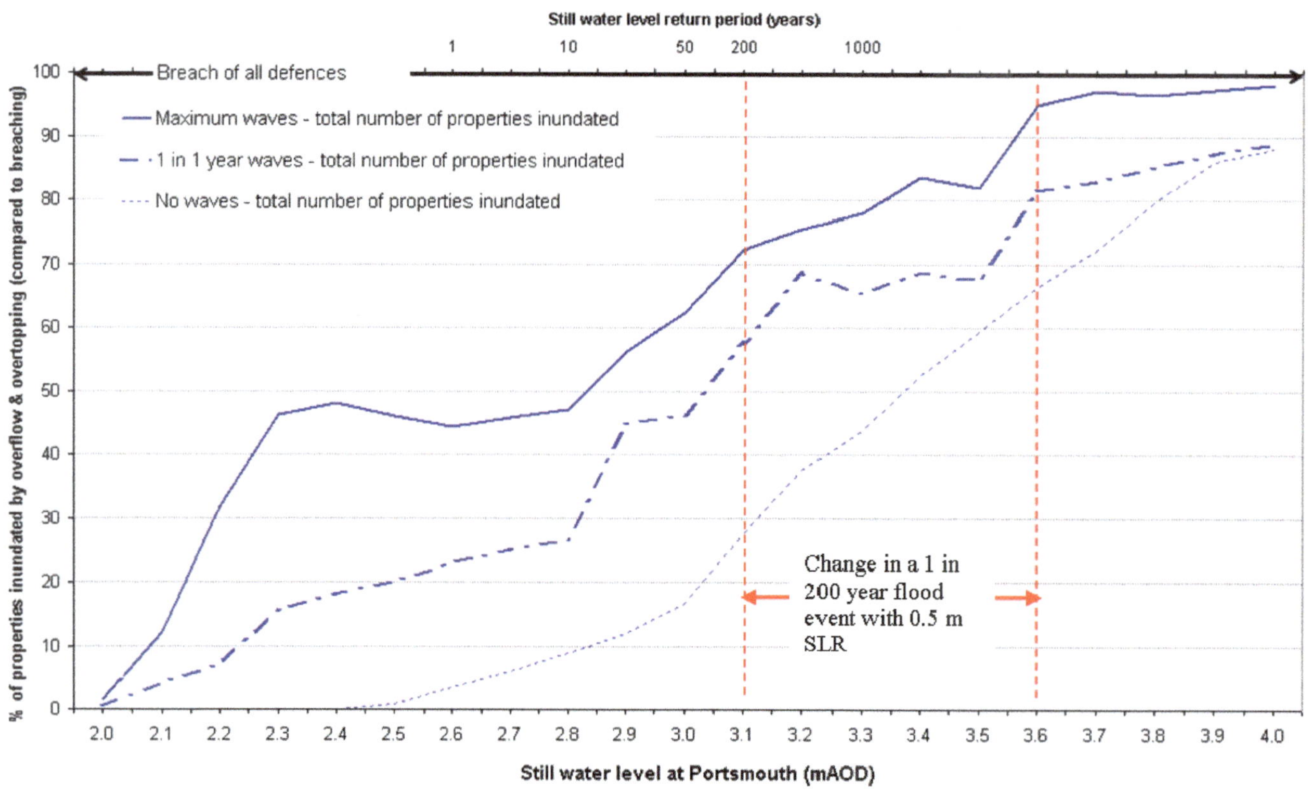

(b)

Figure 9. (**a**) Raw inundation modelling results for properties within modelled coastal flood outlines across the Solent–water level return period refers to annual probability of that water level for 2011; (**b**) Percentage of properties inundated by overflow and overtopping defence failures (compared to breach of all defences).

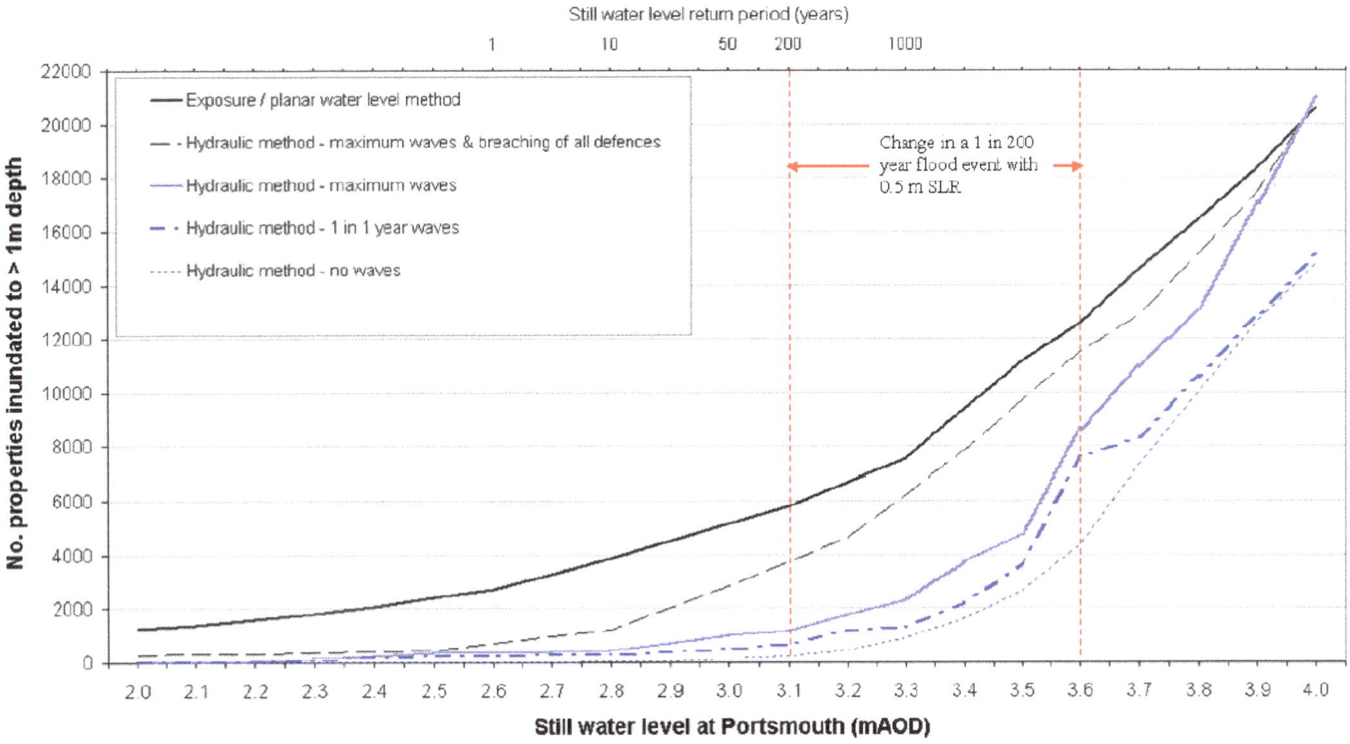

(a)

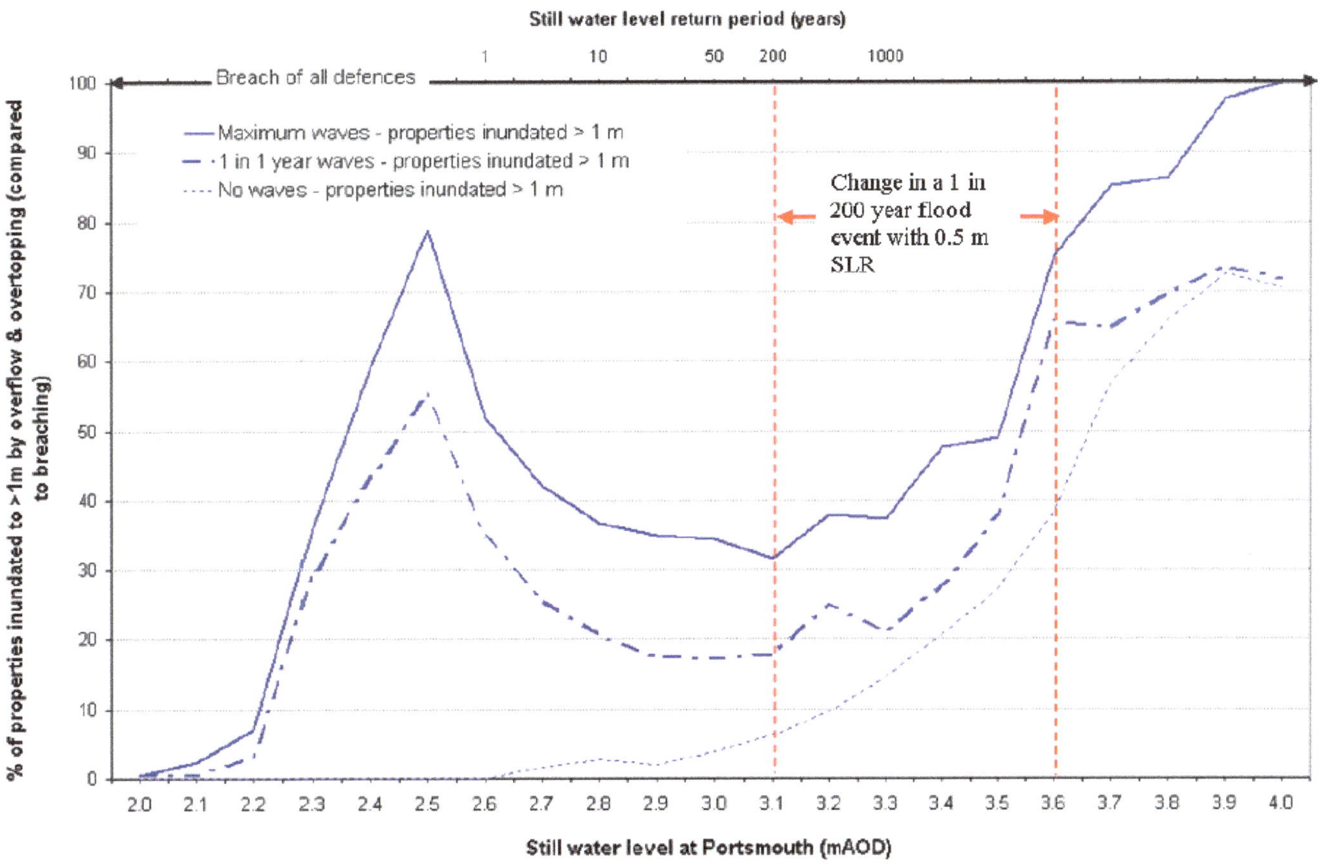

(b)

Figure 10. (**a**) The number of properties in the Solent inundated to a water depth of greater than 1 m; (**b**) Percentage of properties inundated to greater than 1 m water depth by overflow and overtopping defence failures (compared to breach of all defences).

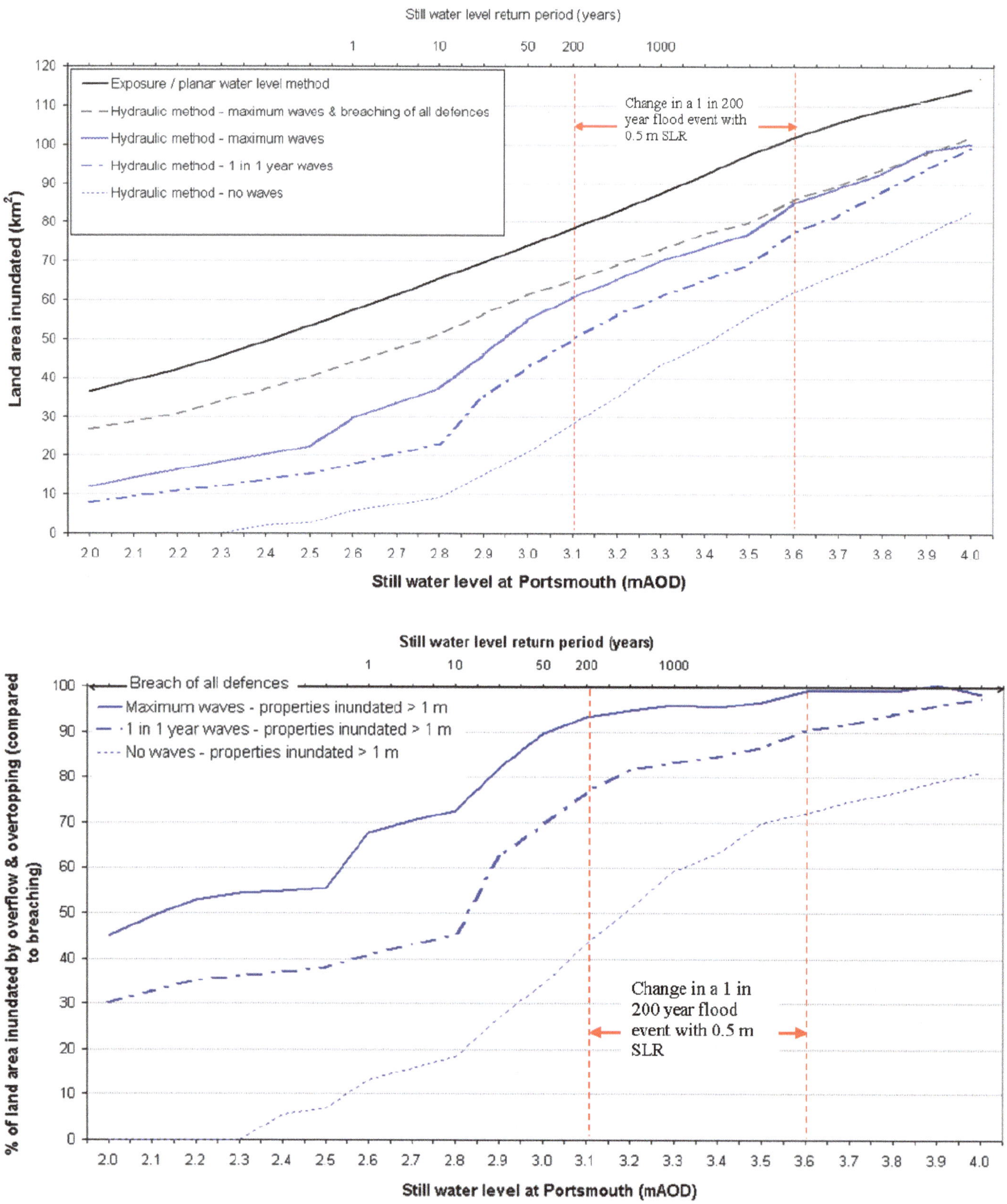

(b)

Figure 11. (**a**) Coastal flood modelling results across the Solent by land area; (**b**) Percentage of land area inundated by overflow and overtopping defence failures (compared to breach of all defences).

UKCP09 [63] global mean sea level projections derived from the IPCC Fourth Assessment Report [93] give an estimated range (5th to 95th percentile) for sea level increase of 18–59 cm between present day (assuming a 1980–1999 baseline) and 2090–2099. For simplicity the effects of a medium UKC09 scenario (of approximately 0.5 m sea-level rise) upon a 1 in 200 year water level flood event are shown in Figure 9, Figure 10, Figure 11. This approximated scenario of 21st century sea-level rise substantially increases the amount of property affected by at an extreme water level coastal flood event in the Solent. For example, in the case of a 1 in 200 year water level flood combined with an annual wave extreme, the total number of buildings in the regionally modelled flood outline increases from 10,128 to almost 24,000 (approximately 140 percent change), although there is a more dramatic 600 percent increase if considering >1 m floods and probably provides a better indication of actual impacts assuming no adaptation (Figure 10a). Even without the effects of waves, a 1 in 200 year flood event in addition to 0.5 m sea-level rise (and no defence improvements) allows nearly 4400 properties to be inundated to greater than 1 m depth.

DISCUSSION

Application of the Methodology

The case study results illustrate how the described methodologies provide a regional overview of the impacts of coastal flood events; and also allow a comparison between the consequences of different types of defence failure. In the region of interest, overtopping is more important than in many previously investigated sites, such as eastern England. This shows that investigating flooding across a range of defence failure mechanisms provides useful insights.

Estimates of coastal flood exposure by the planar water level method were subject to substantial uncertainty, stemming from vertical accuracy of the DEM, compilation of property datasets, and definition of what comprised inundation (*i.e.*, whether the interior or garden of the property is affected). Based upon ±0.15 m variation in the vertical LiDAR accuracy, the regional exposure estimate (generated by the count of buildings in the flood outline) could vary upwards by more than 30 percent, and downwards by approximately 10 percent. The reason for this asymmetry in uncertainty was because the modelled flood outline could be used to select point vectors representing property locations. The upper absolute exposure estimate was different because it also comprised adding a 20 m buffer (an approximation for the spatial footprint of properties within the dataset). Similar principles are applicable to the hydraulic modelling outputs, although accounting for defence failures, mass conservancy and hydraulic connectivity of flood spreading outweigh this effect.

The hydraulic modelling method performed well, using high quality defence and floodplain data that exists in the Solent. The greatest uncertainties in boundary condition processes and resultant inundation was for wave overtopping onto floodplains that are above the peak still water level. In these situations the irregular nature of waves transmitted over a structure and interactions with drainage may be particularly complex to model in a simplified way. The flood modelling method was computationally fast and differentiated from many existing regional approaches by the high level of spatial detail in schematisation of defences and floodplain, with wave overtopping and outflanking flood mechanisms included in all inundation simulations.

Flood Impacts in the Solent

The 1 in 200 year flood simulation demonstrates how the impacts of an extreme present day flood event could be significantly greater than impacts seen on the 10th March 2008. Under equivalent wave loadings, the approximately 30 cm increase in the boundary water levels quadruples the amount of property likely to be affected by inundation (which would increase further with larger waves). The effect of more severe and widespread defence failures (than those seen on the 10th March 2008) would also involve more prolonged , widespread and faster flowing floodplain water; with the 1 in 200 year water level potentially causing sea water to inundate between 30–65 km^2 land. Flood simulations identified significant flooding in populated areas outside of the building dataset; notably at the largest holiday park in the region at Selsey which can accommodate up to 12,500 people. Areas of this site were inundated to depths between 1–2 m in both the annual and extreme overtopping scenarios for the 1 in 200 year simulations. Statistics for people per property and occupancy rates from the 2001 UK Census suggest that on average at least two people may occupy each building in the dataset; hence the population impacted by the large flood events in the region is likely to be substantial and will grow significantly with 21st century sea-level rise (although the defences at Selsey are currently being upgraded). Without changes to defences there is a large increase in the amount of property threatened by flooding when applying the 0.5 m sea-level rise example. Hydraulic modelling of the worst possible 1 in 200 year coastal flood at present-day sea level suggested that approximately 500 properties lay in areas where flood waters may reach 2 m or deeper; this number trebling with the 0.5 m sea-level rise example. Without breach (at present day sea levels) this is far less, although with 0.5 m sea-level rise this level of flooding occurs when a 1 in 200 year event combines with only an annual extremity of wave scenario (without breach failures included). Due to this, and the increasing impacts of overflow and overtopping failures with sea-level rise (shown in Figure 9b, Figure 10b and Figure 11b) the importance of future defence improvements is apparent, or a significant increase in flood consequences is inevitable.

Of the ninety-nine flood compartments in the Solent (Section 3.2), twelve are (at present day sea levels) likely to experience significantly greater flood event impacts due to breaching failures (in comparison to severe overtopping events). These compartments are quite large, and account for 40 percent of the land area that is at risk of flooding in the Solent. However, it is apparent that other failure mechanisms are currently significant to flood impacts of both frequently occurring and extreme events in this region.

Directions for Further Research

This methodology would be complimented if used alongside more spatially descriptive data of regional coastal flood sources, and receptors. The former would be possible by applying nearshore hydrodynamic modelling and the latter by supplementing existing datasets with spatial distributions of population, vulnerability and key infrastructure. The defence and floodplain datasets in the region could be used for higher resolution modelling. Inundation modelling using a coarsened, interpolated DEM has been observed in other studies to overestimate flood extent and underestimate flood depth [49], and modelling of urban inundation benefits significantly from finer discretisation of space [49,70]. This may only be feasible or offer significant benefits at some locations in the region; the most obvious choices being the region's two cities (Portsmouth and Southampton), and low-lying flood compartments where breaching is a threat.

The fast run-time of the regional model and easy integration with GIS could have other applications. For example, within operational real-time coastal flood forecasting systems [94] the modelling of flood pathways (defence failures and inundation mapping) could be better integrated to benefit real-time flood management [31,95,96,97], e.g., by enabling more efficient and strategic flood warning dissemination, improved visual information (to communicate flood event outcomes), and more quantitative appraisal of warning system performance.

The development of practical methods for the application of wave overtopping to the boundaries of coastal inundation models is an important area of future research, for which limited guidance is currently available. Uncertainty analysis is also recognised as crucial to numerical modelling studies [98] and could be more formally integrated for a more comprehensive analysis of present and future flood impacts e.g., [51]. Better calibration is recommended; with compilation of flood data when available to improve models and their interpretation. The 10th March 2008 data (Section 3.4.1) indicated lack of damage to property to flooding below certain water depths, although this also demonstrated the difficulty of using counts of flooded properties to validate models. However, this information probably helped to generate a more realistic interpretation of flood impacts from the synthetic flood event simulations.

CONCLUSIONS

A method for regional analysis of coastal flooding has been demonstrated, using the Solent, UK as a case study. It integrates a number of existing approaches for modelling defence response and floodplain inundation. This allowed individual, synthetic and real, coastal flood events to be simulated, providing a simple evaluation of regional coastal flood impacts. By modelling across different water levels, an overview of the uncertainty relating to boundary condition variations is available. This also showed how overtopping and overflow mechanisms contribute a large present day flood threat across the Solent. Breaching has the potential to generate significant impacts across a range of present day still water levels; although across the range of still water levels expected with 21st century sea-level rise the effects of overflow and overtopping are similar to the effects of broad-scale breaching.

This research offers benefits for the understanding of coastal flooding in a region that has undergone less analysis than the east and west coasts of the UK. This research was also produced in recognition that there is an increasing need to establish methods which improve the understanding of coastal flood events. The run-time for flood event simulations is rapid, allowing multiple uses; such as (1) flood warning; (2) defence prioritisation; (3) floodplain mapping and land-use planning; and (4) operational coastal management. Furthermore outputs are simple and flexible enough to communicate flood impacts to non-scientific audiences. It is recommended that these approaches be further explored, particularly for Portsmouth and Southampton where the majority of coastal population and infrastructure reside, as well as areas where breaching is a threat. Risks and management approaches could change differently for Portsmouth and Southampton over the coming century. For the former, risk is already high and investments in defences already exist with the need for upgrade recognised. Southampton is at present mostly undefended, therefore adaptation will represent a bigger change from current practice, although the needs for a response are increasingly recognised. Globally there are many coasts such as the Solent where surges are not large and flood risk is currently not a major issue; but sea-level rise will significantly exacerbate flood risks. The transferability of this modelling to other coastal regions is in principle quite simple. However, flood predictions were sensitive to small variations in crest height; whilst high resolution and recent datasets (e.g., for defences, floodplains, and extreme sea level analysis) is not available everywhere.

ACKNOWLEDGMENTS

This research was part of a Ph.D funded by the Engineering and Physical Sciences Research Council (EEPSRC) and the 'SemSorGrid4Env' (SG4E) EU research project (http://www.semsorgrid4env.eu/home.jsp). Data was provided by the Channel Coastal Observatory (http:www.channelcoast.org), the Environment Agency Southern Region and the Havant Borough Council/Portsmouth City Council Coastal Defence Partnership. Paul Bates, Jeff Neal and Tim Fewtrell from the University of Bristol provided LISFLOOD-FP and expert guidance.

REFERENCES

1. Van Koningsveld, M.; Mulder, J.P.M.; Stive, M.J.F.; van der Valk, L.; van der Weck, A.W. Living with sea-level rise and climate change: A case study of the netherlands. *J. Coast. Res.* **2008**, *24*, 367–379.
2. Dawson, R.J.; Ball, T.; Werritty, J.; Werritty, A.; Hall, J.W.; Roche, N. Assessing the effectiveness of non-structural flood management measures in the thames estuary under conditions of socio-economic and environmental change. *Glob. Environ. Change* **2011**, *21*, 628–646.
3. Gerritsen, H. What happened in 1953? The big flood in the netherlands in retrospect. *Philos. Trans. R. Soc. A Math. Phys. Eng. Sci.* **2005**, *363*, 1271–1291.
4. Baxter, P.J. The east coast big flood, 31 January–1 February 1953: A summary of the human disaster. *Philos. Trans. R. Soc. A Math. Phys. Eng. Sci.* **2005**, *363*, 1293–1312.
5. McRobie, A.; Spencer, T.; Gerritsen, H. The big flood: North sea storm surge. *Philos. Trans. R. Soc. A Math. Phys. Eng. Sci.* **2005**, *363*, 1263–1270.
6. Bütow, H. *Die große flut in Hamburg: Eine Chronik der Katastrophe vom Februar 1962*; Verlag: Hansestadt, German, 1963.
7. Von Storch, H.; Woth, K.; Gönnert, G. Storm surges—The case of Hamburg, Germany. *ESSP OSC Panel Session on "GEC, Natural Disasters, and Their Implications for Human Security in Coastal Urban Areas"*, Available online: http://www.safecoast.nl/editor/databank/File/hamburg-storms.pdf (accessed on 26 January 2012).
8. Battjes, J.; Gerritsen, H. Coastal modelling for flood defence. *Philos. Trans. R. Soc. Lond. Ser. A Math. Phys. Eng. Sci.* **2002**, *360*, 1461–1475.

9. Bates, P.D.; Dawson, R.J.; Hall, J.W.; Horritt, M.S.; Nicholls, R.J.; Wicks, J.; Mohamed Ahmed Ali Mohamed, H. Simplified two-dimensional numerical modelling of coastal flooding and example applications. *Coast. Eng.* **2005**, *52*, 793–810.

10. Brown, J.D.; Spencer, T.; Moeller, I. Modeling storm surge flooding of an urban area with particular reference to modeling uncertainties: A case study of Canvey Island, United Kingdom. *Water Resour. Res.* **2007**, *43*, W06402.

11. Gallien, T.W.; Schubert, J.E.; Sanders, B.F. Predicting tidal flooding of urbanized embayments: A modeling framework and data requirements. *Coast. Eng.* **2011**, *58*, 567–577.

12. Smith, R.A.E.; Bates, P.D.; Hayes, C. Evaluation of a coastal flood inundation model using hard and soft data. *Environ. Model. Softw.* **2012**, *30*, 35–46.

13. Narayan, S.; Hanson, S.; Nicholls, R.J.; Clarke, D.; Willems, P.; Ntegaka, V.; Monbaliu, J. A holistic assessment of coastal flood systems using system diagrams and the source-pathway-receptor (spr) concept. *J. Nat. Hazards Earth Syst. Sci.* **2012**. in press.

14. Dawson, R.; Peppe, R.; Wang, M. An agent-based model for risk-based flood incident management. *Nat. Hazards* **2011**,*59*, 167–189.

15. RMS. *Hurricane Katrina: Profile of a Super Cat, Lessons and Implications for Catastrophe, Risk Management Solutions; Risk Management Solutions*; Risk Management Solutions, Inc.: Newark, CA, USA, 2005. Available online: http://www.Rms.Com (accessed on 14 June 2011).

16. Hanson, S.; Nicholls, R.; Ranger, N.; Hallegatte, S.; Corfee-Morlot, J.; Herweijer, C.; Chateau, J. A global ranking of port cities with high exposure to climate extremes. *Clim. Change* **2011**, *104*, 89–111.

17. Lumbroso, D.M.; Vinet, F. A comparison of the causes, effects and aftermaths of the coastal flooding of england in 1953 and france in 2010. *Nat. Hazards Earth Syst. Sci.* **2011**, *11*, 2321–2333.

18. Kolen, B.; Slomp, R.; van Balen, W.; Terpstra, T.; Bottema, M.; Nieuwenhuis, S. *Learning from French Experiences with Storm Xynthia–Damages after a Flood*; ©HKV LIJN IN WATER: Waterdienst, The Netherlands, 2010.

19. Muir Wood, R.; Bateman, W. Uncertainties and constraints on breaching and their implications for flood loss estimation. *Philos. Trans. R. Soc. A. Math. Phys. Eng. Sci.* **2005**, *363*, 1423–1430.

20. Haigh, I.; Nicholls, R.; Wells, N. Rising sea levels in the English Channel 1900 to 2100. *Proc. ICE Maritime Engineer.***2011**, *164*, 81–92.

21. Menéndez, M.; Woodworth, P.L. Changes in extreme high water levels based on a quasi-global tide-gauge data set. *J. Geophys. Res.* **2010**, *115*, C10011.

22. Wahl, T.; Jensen, J.; Frank, T.; Haigh, I. Improved estimates of mean sea level changes in the german bight over the last 166 years. *Ocean Dyn.* **2011**, *61*, 701–715.

23. De Moel, H.; van Alphen, J.; Aerts, J.C.J.H. Flood maps in Europe—Methods, availability and use. *Nat. Hazards Earth Syst. Sci.* **2009**, *9*, 289–301.

24. Reeve, D.E.; Burgess, K.A. A method for the assesment of coastal flood risk. *IMA J. Manag. Math.* **1993**, *5*, 197–209.

25. Hall, J.; Dawson, R.J.; Sayers, P.B.; Rosu, C.; Chatterton, J.B.; Deakin, R.A. A methodology for national-scale flood risk assessment. *Water Marit. Eng. ICE* **2003**, *156*, 235–247.

26. Dawson, R.; Hall, J. Adaptive importance sampling for risk analysis of complex infrastructure systems. *Proc. R. Soc. A Math. Phys. Eng. Sci.* **2006**, *462*, 3343–3362.

27. Gouldby, B.; Sayers, P.; Mulet-Marti, J.; Hassan, M.; Benwell, D. A methodology for regional-scale flood risk assessment. *Water Manag.* **2008**, *161*, 169–182.

28. EA. *Flooding in England: A National Assessment of Flood Risk*; Environment Agency: Rotherham, UK, 2009. Available online: http://publications.Environment-agency.Gov.Uk/pdf/geho0609bqds-e-e.Pdf (accessed on 23 April 2012).

29. Defra. *Shoreline Management Plan Guidance Volume 1: Aims and Requirements*; DEFRA: London, UK, 2006. Available online: http://www.Defra.Gov.Uk/publications/files/pb11726-smpg-vol1–060308.Pdf (accessed on 25 January 2012).

30. Motyka, J.M.; Brampton, A.H. *Coastal Management: Mapping of Littoral Cells Ref: SR328*; HR Wallingford: Oxfordshire, UK, 1993.

31. DEFRA; Environment Agency. *Best Practice in Coastal Flood Forecasting: R&d Technical Report FD2206/TR1; HR Wallingford Report TR 132*; DEFRA: London, UK, 2004.

32. Bradbury, A.; Kidd, R. Hurst spit stabilisation scheme: Design and construction of beach recharge. In 33rd MAFF Conference of River and Coastal Engineers, Keele University, Staffordshire, UK, 1–3 July 1998.

33. Haigh, I.; Butcher, P.; Harris, J.M.; Cooper, N.J.; Trip, I. Developing an improved understanding of storm surges in the solent. In Proceedings of the39th DEFRA Flood and Coastal Management Conference, York, UK, 29 June–1 July 2004.

34. Law, C.R. Storm surges in the English Channel. *Hydrogr. J.* **1975**, *2*, 30–34.

35. Haigh, I. Extreme Sea Levels in the English Channel. Ph.D Dissertation, University of Southampton, Hampshire, UK, 2009.

36. Ruocco, A.; Nicholls, R.; Haigh, I.; Wadey, M. Reconstructing coastal flood occurrence combining sea level and media sources: A case study of the Solent, UK since 1935. *Nat. Hazards* **2011**, *59*, 1773–1796.

37. Pugh, D.T. Tides, surges and mean sea-level. In *A Handbook for Engineers and Scientists*; Wiley: Chichester, UK; Volume 1987.

38. Wells, N.C.; Baldwin, D.J.; Wang, J.Y.; Collins, M.B. Modelling of extreme storm surge events in the English Channel for the period 14–18 December 1989. *Glob. Atmos. Ocean Syst.* **2001**, *7*, 275–294.

39. Mason, T.; Bradbury, A.; Poate, T.; Newman, R. Nearshore wave climate of the English Channel—Evidence for

bi-modal seas. In *31st International Conference on Coastal Engineering, Hamburg, Germany, 31 Auguest–5 September 2008*; Jane, M.S., Ed.; World Scientific: Singapore; Hamburg, Germany, 2009; pp. 605–616.

40. Evans, E. *Foresight: Scientific Summary. Volume ii, Managing Future Risks*; Office of Science and Technology: London, UK, 2004.

41. NFDC. *Cabinet 06/07/09 Report A: North Solent Shoreline Management Plan*; New Forest District Council: Hampshire, UK, 6 July 2009. Available online: http://www.Newforest.Gov.Uk/committeedocs/cab/cdr05054.Pdf (accessed on 15 December 2011).

42. Portsmouth City Council (PCC). *Portsmouth City Local Plan, First Review 2001–2011. Supplementary Planning Guidance: Flood Protection. Approved by the Executive Member for Planning, Regeneration, Economic Development and Property 6th February 2004*; Portsmouth City Council: Wales, UK, 2004. Available online: http://www.portsmouth.gov.uk/media/Flood_Protection_SPG.pdf (accessed on 26 August 2009).

43. RIBA; ICE. *Facing up to Rising Sea-Levels: Retreat? Defend? Attack? The Future of Our Coastal and Estuarine Cities; Royal Institute for British Architects*; Institute of Civil Engineers: London, UK, 2009. Available online: http://www.buildingfutures.org.uk./downloads/Facing_Up_To_Rising_Sea_Levels.pdf (accessed on 31 January 2010).

44. Atkins. *Strategic Flood Risk Assessment for the Partnership for Urban South Hampshire; Document Number: 5049258/72/DG/048*; Atkins limited: London, UK, 2007. Available online: http://www.environment-agency.gov.uk/static/documents/Research/PUSH_SFRA.pdf (accessed on 24 April 2012).

45. OS. *Ordnance Survey Land-Form Profile™ User Guide; V4.0–5/2001© Crown Copyright*; Ordnance Survey: Southampton, UK, 2001. Available online: http://www.ordnancesurvey.co.uk (Accessed on 6 July 2011).

46. CCO. *Strategic Regional Coastal Monitoring Programmes*; Channel Coastal Observatory: Southampton, UK, 2011. Available online: http://www.channelcoast.org (accessed on 23 April 2012).

47. Xharde, R.; Long, B.; Forbes, D. Accuracy and limitations of AIrborne Lidar surveys in coastal environment. In Proceedings of theGeoscience and Remote Sensing Symposium, 2006; IGARSS 2006. IEEE International Conference, Denver, CO, USA, 31 July–4 August 2006; pp. 2412–2415, ISBN: 0-7803-9510-7.

48. Webster, T.L.; Forbes, D.L.; MacKinnon, E.; Roberts, D. Flood-risk mapping for storm-surge events and sea-level rise using lidar for southeast new brunswick. *Can. J. Remote Sens.* **2006**, *32*, 194–211.

49. Fewtrell, T.J.; Bates, P.D.; Horritt, M.; Hunter, N.M. Evaluating the effect of scale in flood inundation modelling in urban environments. *Hydrol. Process.* **2008**, *22*, 5107–5118.

50. De Mers, M.N. *Fundamentals of Geographic Information Systems*, 4th ed; John Wiley Sons Inc.: New York, NY, USA, 2009.

51. Purvis, M.J.; Bates, P.D.; Hayes, C.M. A probabilistic methodology to estimate future coastal flood risk due to sea level rise. *Coast. Eng.* **2008**, *55*, 1062–1073.

52. Dixon, M.J.; Tawn, J.A. *Extreme Sea-Levels at the UK a-Class Sites: Site-by-Site Analysis; Pol Internal Document Number 65*; Proudman Oceanographic Laboratory: Liverpool, UK, 1994. Available online: http://www.pol.ac.uk/ntslf/pdf/id65.pdf(accessed on 23 April 2012).

53. Dixon, M.J.; Tawn, J.A. *Extreme Sea-Levels at the UK a-Class Sites: Optimal Site-Bysite Analysis and Spatial Analyses for the East Coast; Pol Internal Document Number 72*; Proudman Oceanographic Laboratory: Liverpool, UK, 1995; p. 298.

54. Dixon, M.J.; Tawn, J.A. *Estimates of Extreme Sea Conditions—Final Report, Spatial Analysis for the UK Coast; Pol Internal Document Number 112*; Proudman Oceanographic Laboratory: Liverpool, UK, 1997; p. 217.

55. Haigh, I.D.; Nicholls, R.; Wells, N. A comparison of the main methods for estimating probabilities of extreme still water levels. *Coast. Eng.* **2010**, *57*, 838–849.

56. Flood Risk Mapping and Data Management Team. *Environment Agency Extreme Water Levels: Bournemouth to Littlehampton*; EA: Rotherham, UK, 2007.

57. EA. *Coastal Flood Boundary Conditions for UK Mainland and Islands; Project: Sc060064/tr2: Design sea Levels*; EA: Bristol, UK, 2011; p. 142.

58. Haigh, I. Quantification of sea-level rise at the Southampton and Portsmouth tide gauges. Personal Communication.

59. NFDC. *North Solent Shoreline Management Plan, Appendix c Baseline Process Understanding, c.1 Assessment of Shoreline Dynamics*; New Forest District Council: Hampshire, UK, 2010. Available online: http://www.newforest.gov.uk/media/adobe/o/s/C1_Coastal_Processes _Literature_Review_%28final_draft%29.pdf (accessed on 30 June 2011).

60. CERC. *Coastal Engineering Research Center 1984: Shore Protection Manual*; Department of the Army, Waterways Experiment Station, Corps of Engineers, Coastal Engineering Research Center, Vicksburg, United States Government Printing Office: Washington, DC, USA, 1984; p. 652.

61. Dawson, R.; Hall, J. Probabilistic condition characterisation of coastal structures using imprecise information. In Proceedings of the 28th International Conference Coastal Engineering Cardiff, UK, 8–12 July 2002; McKee-Smith, J., Ed.; World Scientific: Hackensack, NJ, USA; Cardiff, UK, 2003; pp. 2348–2359.

62. Flikweert, J.; Simm, J. Improving performance targets for flood defence assets. *J. Flood Risk Manag.* **2008**, *1*, 201–212.

63. Lowe, J.A.; Howard, T.; Pardaens, A.; Tinker, J.; Holt, J.; Wakelin, S.; Glenn, M.; Leake, J.; Wolf, J.; Horsburgh, K.; *et al.UK Climate Projections Science Report: Marine and Coastal Projections*; Met Office Hadley Centre:

Exeter, UK, 2009; p. 99.

64. Bates, P.D.; Horritt, M.S.; Fewtrell, T.J. A simple inertial formulation of the shallow water equations for efficient two-dimensional flood inundation modelling. *J. Hydrol.* **2010**, *387*, 33–45.

65. Bates, P.D.; de Roo, A.P.J. A simple raster-based model for flood inundation simulation. *J. Hydrol.* **2000**, *236*, 54–77.

66. Hunter, N.M.; Horritt, M.S.; Bates, P.D.; Wilson, M.D.; Werner, M.G.F. An adaptive time step solution for raster-based storage cell modelling of floodplain inundation. *Adv. Water Resour.* **2005**, *28*, 975–991.

67. Dawson, R.J.; Hall, J.W.; Bates, P.D.; Nicholls, R.J. Quantified analysis of the probability of flooding in the thames estuary under imaginable worst-case sea level rise scenarios. *Int. J. Water Resour. Dev.* **2005**, *21*, 577–591.

68. Hunter, N.M.; Bates, P.D.; Horritt, M.S.; Wilson, M.D. Simple spatially-distributed models for predicting flood inundation: A review. *Geomorphology* **2007**, *90*, 208–225.

69. Fewtrell, T.J.; Duncan, A.; Sampson, C.C.; Neal, J.C.; Bates, P.D. Benchmarking urban flood models of varying complexity and scale using high resolution terrestrial lidar data. *Phys. Chem. Earth Parts A/B/C* **2011**, *36*, 281–291.

70. Hunter, N.M.; Bates, P.D.; Neelz, S.; Pender, G.; Villanueva, I.; Wright, N.G.; Liang, D.; Falconer, R.A.; Lin, B.; Waller, S.; *et al.* Benchmarking 2D hydraulic models for urban flooding. *Proc. Inst. Civil Engineers Water Manag.* **2008**, *161*, 13–30.

71. Pullen, T.; Allsop, N.W.H.; Bruce, T.; Kortenhaus, A.; Schuttrumpf, H.; van der Meer, J. *Eurotop, Wave Overtopping of Sea Defences and Related Structures—Assessment Manual*, 2007. Available online: www.overtopping-manual.com(accessed on 24 April 2012).

72. Owen, M.W. *Design of Seawalls Allowing for Wave Overtopping; Report Ex 924*; HR Wallingford: Oxfordshire, UK, 1980.

73. *Technical Report—Wave Run-Up and Wave Overtopping at Dikes*; TAW: Delft, The Netherlands, 2002.

74. Gouldby, B.; Kingston, G.; Wills, M.; Van Gelder, P.; Buijis, F.; Kortenhaus, A. *Flood Defence Reliability Calculator: Reliable User Manual; Floodsite Report Number: T07–10–08*; January 2008.

75. Allsop, W.; Kortenhaus, A.; Morris, M. *Failure Mechanisms for Flood Defence Structures; Floodsite Report Number: T04–06–01*; HR Wallingford Ltd.: Oxfordshire, UK, 2007.

76. Vrouwenvelder, A.C.W.M.; Steenbergen, H.M.G.M.; Slijkhuis, K.A.H. Theoretical Manual of Pc-Ring, Part B: Statistical Models (in dutch); No. 98-con-r1431. Delft, The Netherland, 2001.

77. Terzaghi, K.; Peck, R.B. *Soil Mechanics in Engineering Practice*; John Wiley & Sons: New York, NY, USA, 1967; p. 729.

78. Powell, K.A. *Predicting the Short Term Profile Response for Shingle Beaches; Hydraulics Research Report sr219*; HR Wallingford: Oxfordshire, UK, 1990.

79. DEFRA; EA, *Performance and Reliability of Flood and Coastal Defences; R&d Technical Report Fd2318/tr 2*; Buijis, F., Simm, J., Wallis, M., Sayers, P., Eds.; DEFRA: London, UK, 2005.

80. Morris, M.; Hassan, M.; Kortenhaus, A.; Visser, P. *Breaching Processes: A State of the Art Review; Floodsite Report Number t06–06–03, Revision Number 5_2_p01*; FLOODsite Consortium: Oxfordshire, UK, 2009.

81. Morris, M.; Hanson, G.; Hassan, M. Improving the accuracy of breach modelling: Why are we not progressing faster?*J. Flood Risk Manag.* **2008**, *1*, 150–161.

82. Wadey, M.; Nicholls, R.J.; Haigh, I.; Hutton, C. Creating a coastal flood event dataset for the solent region-the 10th March 2008 storm surge event. 2012. in preparation.

83. Cope, S. Details of the breach of the medmerry/selsey shingle barrier during the storm on 10 March 2008. Personal Communication, New Forest District Council/Channel Coastal Observatory: Southampton, UK.

84. EA, Solent & South Downs Area Report 10 March Tidal Flooding Event, Hampshire & Isle of Wight: A report of events, Consequences and Conclusions, Environment Agency, Rotherham, UK, 2010. Version 2.

85. Wadey, M.P. Photograph taken by Matthew Wadey from the Channel Coastal Observatory overlooking Southampton Docks. National Oceanography Centre, Southampton, UK, 2:05PM, 10 March 2008.

86. Evacuations as waves flood park. *BBC World News*, 10 March 2008. Available online: http://news.bbc.co.uk/1/hi/england/sussex/7287880.stm (accessed on 25 January 2012).

87. Emsworth Residents Association. Flooding in Emsworth, Borough of Havant, during a Storm Surge in March 2008; Emsworth_Flood_2008_1.jpg. 10 March 2008. Available online: http://www.floods.emsworthhants.org.uk/ (accessed on 24 April 2012).

88. NRA, Lymington/Pennington Flood Investigation–Interim Report; Report for National Rivers Authority Southern Region; Ea Report Reference Number: IMP/sch/008/5054, Southern Projects Ltd. Jersey, UK, 1990.

89. Stripling, S.; Bradbury, A.P.; Cope, S.N.; Brampton, A.H. *Understanding Barrier Beaches; R&D Technical Report Fd1925/Tr*; JOINT Defra/ea Flood and Coastal Erosion Risk Management R&D Programme: London, UK, 2008. Available online: http://sciencesearch.defra.gov.uk (accessed on 24 April 2012).

90. Nicholls, R. The Stability of the Shingle Beaches in the Eastern Half of Christchurch Bay. Ph.D Dissertation, University of Southampton, Hampshire, UK, 1985.

91. Nicholls, R.J.; Webber, N.B. The past, present and future evolution of hurst castle spit, Hampshir. *Prog. Oceanogr.* **1987**,*18*, 119–137.

92. Jonkman, S.N.; Penning-Rowsell, E. Human instability in flood flows. *JAWRA J. Am. Water Resour. Assoc.* **2008**, *44*, 1208–1218.

93. Pachauri, R.K., Reisinger, A., Eds.; *Contribution of*

Working Groups I, II and III to the Fourth Assessment Report of the Intergovernmental Panel on Climate Change; IPCC, Climate Change 2007: Synthesis Report; IPCC: IPCC, Geneva, Switzerland, 2007. Available online: http://www.ipcc.ch/publications_and_data/ar4/syr/en/contents.html (accessed on 24 April 2012).

94. Flather, R.A. Existing operational oceanography. *Coast. Eng.* **2000**, *41*, 13–40.

95. Ray, J.; Coulling, E.; Evans, S.Y.; Spink, K. Developing a practical tool for coastal flood forecasting and warning. ICE Coastal Management: Innovative Coastal Zone Management.Sustainable Engineering for a Dynamic Coast, Belfast, UK, 2011.

96. Wadey, M.; Nicholls, R.; Hutton, C. Threat of coastal inundation in the solent: Real-time forecasting. ICE Coastal Management: Innovative Coastal Zone Management. Sustainable Engineering for a Dynamic Coast, Belfast, UK, 2011.

97. DEFRA; EA, *Improving Flood Warnings: Final Report: Improving Institutional and Social Responses to Flooding; Science Report: Sc060019-Work Package 1a*; Twigger-ross, C., Fernández-bilbao, A., Tapsell, S., Walker, G., Watson, N., Collaborators, C., Rose, P., Dade, H., Deeming, E., Eds.; DEFRA: London, UK, 2009; p. 67, ISBN: 978-1-84911-058-7.

98. Pappenberger, F.; Beven, K.J. Ignorance is bliss: Or seven reasons not to use uncertainty analysis. *Water Resour. Res.***2006**, *42*, W05302.

INDEX

A

Acute Respiratory Disease (ARD) 97
Advanced Very High Resolution Radiometer (AVHRR) 81
Analysis of principal component (ACP) 52, 53
Atomic Absorption Spectrometry (AAS) 33

C

Chart datum (CD) 86
Coastal area system 21
Coastal ecosystems 75, 76, 77
Coastal environment 7, 13
Coastal environments 6, 7, 8, 9, 11, 12, 13
Coastal Erosion 1, 2, 4
Coastal flooding 112, 113, 114, 119, 121, 126, 127
Coastal floods 113
Coastal geomorphology 15, 16
Coastal hydrodynamics 16
Coastal management 19, 20, 21, 22, 26, 27, 28, 29
Coastal Protection 55, 56
Coastal sedimentary 15
Coastal tourism 97
Coastal vulnerability 86, 92, 95
Coastal vulnerability index 88
Coastal vulnerability index (CVI) 85, 86
Coastal zones 5, 6, 8, 9, 11, 12, 13
coastline 69, 70
Comparative analysis 32
Cultural landscape 16

D

Digital elevation model (DEM) 81
Digital Elevation Models (DEM's). 81
Digital terrain model (DTM) 114

E

Ecosystem 5, 12, 103, 104, 105, 108, 109, 110
Environmental protection 31
Environmental remediation 71, 73

F

Fisheries management 100
Flood consequences 112, 125
Flood simulation 121
Forming interconnected systems 75
Foundation for Environmental Education (FEE) 97

G

Gastroenteritis (GI) 97
Geographical information systems (GIS) 114
Geomorphological evolution 16

H

Homogeneous distribution 27
Homogenization 33
Hydraulic modelling method 114, 125
Hydrodynamic control 15
Hydrodynamics 16
Hydrolectrical dams 104

I

Integrated Coastal Zone Management 27, 29
Integrated Coastal Zone Management (ICZM) 55
Integrated management 102
Investigation 70

L

Land occupation 16
Land Use/Land Cover (LULC) 103
Lebanese coast 31, 32, 53
Lymington, 119

M

Magnitude 6, 7, 8, 9, 11, 12
Mangrove 79, 80, 81, 82, 84
Mangroves 79, 80, 81, 83, 84
Marine ecosystems 101
Marine environment 101

Marine fish 58
Marine pollution 69, 70
Marine water 58
Mean Sea Level (MSL) 86
Mental wellbeing 5, 13
Millennium Ecosystem Assessment 103, 108, 109
Motivations 101

N

Natural environment 5, 6, 8, 9, 11, 12
Natural landscapes 16
Non Governmental Organization (NGO) 97
Normalized Difference Water Index (NDWI) 88

P

Physical Intervention (NPI) 56
Presidency of Meteorology and Environment 81

Q

Quantitative distribution 61

R

Rapid sedimentation 59
Real-time kinematic (RTK) 115
Reconstruction 16
Regional analysis 126
Regional management fishery organisation 99
Rich biodiversity 101
road construction 20
Rural 16

S

Scientists 76
Sea surface salinity (SSS) 76
Shoreline 1, 2, 4
Smoothed Particle Hydrodynamics Method (SPH) 3
Statistical analysis 60
Supervising sea surface temperature (SST) 76
Support vector machine (SVM) 88
SWOT analysis 21, 22, 29
Synthetic Aperture Radar (SAR) 3

T

Threats Analysis 21
Tourism infrastructure 103, 108

U

United Nations Environment Programme (UNEP) 85
Urbanization 19, 28, 29

V

Velocity 3